LC (8) GAN

Bootstrapping Machine Learning

Website

www.louisdorard.com/bml-paperback

Release

Version 1.0.5 — 8 April 2014

Written by

Louis Dorard

ELECTRONIC VERSION AND RESOURCES

Welcome to the first edition of Bootstrapping Machine Learning!

What you're reading was initially designed as an ebook and it was then adapted for print. If you want to get an electronic version of this book, please forward your receipt or proof of purchase to bml-ebook@codole.com. You will then receive instructions to get DRM-free PDF, ePub and Mobi versions of the book.

Bootstrapping Machine Learning also comes with additional resources to use on your computer: videos, screencasts, IPython notebooks, code, sample datasets, and a virtual machine. To learn more, go to:

http://www.louisdorard.com/bml-paperback

CONTENTS

Chapter 1: Introduction 7
Who is this book for? 8
Structure of this book 11

Chapter 2: Machine Learning and Artificial Intelligence 13
Motivations 14
ML-powered apps 18
A graphical example 22
Learning vs hand-coding the rules 29
When ML fails 33

Chapter 3: Concepts 41
Types of ML problems 42
The two pillars of ML 45
ML-as-a-Service 51

Chapter 4: Examples..59
Format 60
Business 63
Apps 77

Chapter 5: Applying Machine Learning to your domain 97
Specifying the right problem 98
Feature engineering 109
Preparing the data 117
Measuring performance 130

Chapter 6: Prediction APIs..141
Which APIs? 142
Specialized Prediction APIs 143
BigML 148
Google Prediction API 155
Other generic Prediction APIs 161

Chapter 7: Case study - Priority Inbox 163
Specifying the problem 164
Feature engineering 167
Preparing the data 173
Results 185

Chapter 8: Wrap-up 193
Conclusions 194
About Louis Dorard 197
Acknowledgments 199
Boring notice 201
Notes 203

CHAPTER 1: INTRODUCTION

1

WHO IS THIS BOOK FOR?

This book is for computer engineers, scientists, hackers, programmers, analysts, and thinkers. We are building businesses and applications, for ourselves or for others. Despite coming from different backgrounds, part of our jobs is to process information. We collect it, we move it from one place to another, and we store it as *data*. The next step after gathering this data is to use our machines to analyze it, learn from it, and create new information based on what they have learned.

This new information might be answers to specific questions we've asked about the data, such as "How much?" and "Which one?" For example, you might record real estate transaction prices and ask the price of a property that has not been sold yet. Or you may have received comments to a blog post and would like to identify which are spam or offensive, to filter them out. Because the answers to these questions are not to be found directly in the data, we call them predictions.

The ability to make such predictions holds a lot of value. When it's built into applications to make them smarter, the users have an improved experience. Observing a user's behavior and gathering data from his interactions with the app can be used to understand his interests and predict his next move. It's like having personalized virtual assistants for each user who work extremely fast to make the app more engaging. Predictions can also be used to improve a business by analyzing and making sense of customer data. One example of that is predicting which customers are likely to cancel a service subscription, and taking appropriate action to keep them from leaving. In all of these examples, making predictions from data results in increased revenue.

Up until a few years ago, the science of making such predictions by having a machine "learn" from data required a strong background in mathematics, probability, statistics, computer science, and a lot of practical experience. Your only chance was to hire experts in the field, or be one yourself. Luckily, the 2010s have seen a major technological advance through the introduction of new Prediction Application Programming Interfaces on the web (Prediction APIs). Prediction APIs can be used with no prior knowledge of data science. Indeed, they simply consist in two methods: one method to feed data, and one method to ask questions and get predictions. My decision to write this book was inspired by everything Prediction APIs can offer.

You will find many textbooks on the market to teach you the inner workings of the Machine Learning algorithms and of the Data Mining techniques that are used to make data-based predictions. They will teach you everything from the theory behind these algorithms to how to implement them and how to apply them to real-world problems. My book is not one of them.

My intent is to cover only the basic concepts so you can start making predictions from data and bootstrap the use of Machine Learning in your projects, applications, or business. I will teach you just enough for you to use Prediction APIs effectively. I will explain how to formulate your problem as a Machine Learning problem, and what the algorithms behind the APIs are supposed to do — not how they work. If you ever decide in the end that you want to become an expert in Machine Learning, I believe you will be in a much better position to do so after reading this book.

I am writing this book to help you make better applications and businesses. I want to show you that these days, everyone is in a position to exploit the power of Machine Learning with minimal coding experience and the use of Prediction APIs. It doesn't take a Master's degree and it doesn't require a big budget. You will actually save a lot of money by not hiring an expert and you will even have better chances to incorporate your domain knowledge by handling things yourself.

I am writing this book for bootstrappers who understand why you should do a job yourself before hiring someone else to do it. I am writing for Über-builders who know how important it is to master all the aspects of a product. I am writing for developers who want to extend their skill-set and for thinkers who want to broaden their horizons. I am writing to help you succeed and differentiate yourself by using Machine Learning.

STRUCTURE OF THIS BOOK

Let's start our journey into the world of Machine Learning and Prediction APIs with some prerequisites. First, I'll situate ML within what we call Artificial Intelligence and other techniques used to make computers "intelligent." I'll continue with some basic concepts and terminology so we can characterize what ML does more precisely and introduce the Prediction APIs that give access to ML as a Service. ML is about learning from examples, so I'll give you examples of ML uses to help you learn the concepts of what ML does. If things ever start to get too abstract for you while reading this book, you can refer to these examples for context.

The rest of the book will focus on guiding you to put Prediction APIs to use. We will see how to formulate your problem as an ML problem, how to prepare your data, and how to integrate predictions in your application. We will then review Prediction APIs and see how to use them in more detail. Be ready to get your hands dirty after that, because we'll be applying everything we said and we'll be using Prediction APIs on a concrete case.

Happy reading!

CHAPTER 2: MACHINE LEARNING AND ARTIFICIAL INTELLIGENCE

2

MOTIVATIONS

Computers are really good at processing lots of information. Think about the millions of messages flowing to the Twitter servers, which then have to figure out trending topics in real time. But they lack intelligence. They need to be programmed with precise lists of basic instructions to perform. This makes it challenging for them to do certain things that humans do very easily, but for which we do not know the exact process. For example, think about recognizing hand-written digits: can you explain to me, step-by-step, how you can recognize numbers that form the ZIP code on a hand-written address?

There are diverse techniques to make computers do "intelligent" or "human-like" things, and they form the field of Artificial Intelligence (AI). We'll have a word of caution, first, on this terminology. "Intelligence" and "human-like" do not mean that computers think and have the same mental capabilities as us humans. It just means that they do things that are non-trivial and that only we could do before, such as playing chess. The AI techniques used in chess programs work by examining possible sequences of moves (player A plays this move, then player B may play these moves, then player A may play those moves, etc.) and by evaluating who has the advantage at the end of each of these sequences, simply by counting who has more pieces on the board. There's not much intelligence in that technique, but it does the job! Chess is an example of something that's easy for computers but hard for humans, but it doesn't mean that computers are more intelligent than us. Another example is recognizing people in video footage: this is *very hard* for computers but dead easy for humans.

Machine Learning (ML) is a set of AI techniques where intelligence is built by referring to examples. In the hand-written digit recognition problem, you would use ML by providing examples of hand-written digits (in the form of scanned images) along with their corresponding integers between 0 and 9. These examples are called the "data." When given a new image of a hand-written digit that's never been seen before, the idea is to compare it to the example images and make a prediction based on the integers that correspond to the examples most similar to the new image. This is very similar to what a human would do; Machine Learning is about making machines reproduce how humans learn.

If computers can be just as smart as a human being on a given task, the immediate benefit is that this task may be performed with greater speed and at lower costs. For certain tasks, our own human reasoning may be, just as in Machine Learning, to compare to examples — or in other words, refer to our own previous experience. In this case, we might even expect computers to outsmart us, thanks to their ability to process more information.

An example of referring to experience is establishing diagnoses based on patients' symptoms. Doctors refer to their own experience or to the collective experience that has been synthesized in the literature. They refer to examples of previous patients for whom a diagnosis was made, and they compare new patients to them. The hope with computers is that their processing power would allow them to consider many more previous patients, but also be able to consider more information about each patient (demographic, diet, clinical measurements, and medical history). Doing so should improve the overall accuracy of predictions.

Machine Learning is a set of AI techniques where intelligence is built by referring to examples.

ML-POWERED APPS

Let me now give a few examples of how ML is used in popular web and mobile apps.

On the web, giant tech companies such as Google, Yahoo and Microsoft use Machine Learning extensively to perform tasks that require intelligence on a huge scale. They are able to show personalized content to each of the hundreds of millions of visitors of their web properties by predicting what is most likely to engage each user. Every time a user interacts with their website, it gives them example data of what content interests which user. This content can be articles on a news site homepage or ads on a search engine results page.

Google Mail (Gmail) is an interesting example of a web application which uses Machine Learning to offer killer new features to its users. With Priority Inbox, Google have taken email categorization a step further than traditional email clients who are able to identify spam emails: Gmail can also identify important emails. I personally find it a huge time saver as it helps me filter the noise in my inbox and keep it clean. Besides, I can help the system learn what is important to *me* by marking or unmarking emails as important. Unlike desktop clients, Gmail is continuously learning and updating in real time with the data collected from all the users' interactions. As you can see, Machine

Learning opens up new ways for users to interact with web applications, with the promise of improving their experience. ML-powered features act as individual assistants to applications' users, by making intelligent predictions based on collected data.

On mobile, Apple has empowered the owners of their iDevices with an actual virtual assistant, Siri. Queries are made to it via voice. The user holds down a button, speaks, and waits for Siri to process the query and respond. The type of queries that Siri can process are making a call, sending a text, finding a business, getting directions, scheduling reminders and meetings, searching the web, and more. In Siri, ML is used all over the place for speech recognition (transforming the audio recording of the query to text), for recognizing the type of query among all those that can be processed, and for extracting the "parameters" of the query (e.g. for a query of the type "reminder", extract the name of the task and time to be reminded at — "remind me to do *a certain task* at *a certain time*"). Once this is done, Siri calls the appropriate service/API with the right parameters, and voilà! The aim here is to allow mobile users to do more with their devices, with less interaction. Often times, mobile usage situations make it more difficult to interact with the device than with a desktop machine. Using Machine Learning is a way to make it easier to interact with mobile apps.

The photo upload interface on the Facebook mobile app gives us an example of how ML can help improve the usage of an app. Immediately after selecting a photo, the app scans the image for faces, and little dialogs are positioned below every face that has been detected, prompting the user to tag the person whose face appears on the picture. This creates more user engagement and allows Facebook to gently remind its users that their experience is more fun if they tag friends on their pictures.

You will find these examples presented in more details in Chapter 4, but bear with me for a moment until I introduce the main concepts of Machine Learning and some terminology (Chapter 3).

ML-powered features act as individual assistants to applications' users, by making intelligent predictions based on collected user data.

A GRAPHICAL EXAMPLE

Let's now have a look at how ML works with a graphical illustration of the intuition behind a categorization algorithm. Consider objects that are characterized by two attributes, and to which you want to assign one of two categories. For instance, let's say you want to categorize iris flower species (this is the "hello world" of ML). We're not botanists, but we can measure petal lengths and widths to characterize the flowers.

Suppose we have collected example data, consisting of records of petal length and width of *setosa* and *virginica* iris flowers. We can plot these measurements as points on a 2D graph showing the petal length on the horizontal axis and the width on the vertical axis. The species information determines the color of each plotted point: orange for virginica and blue for setosa.

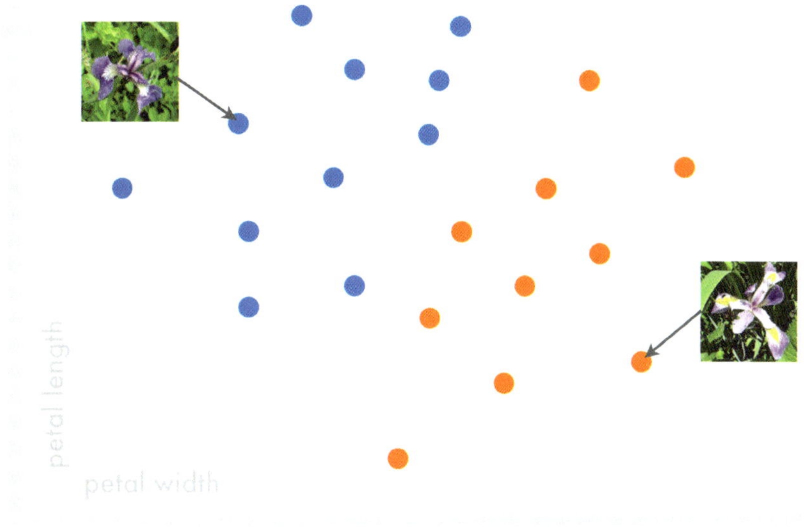

Now imagine that we're given new pairs of length-width petal measurements, but we're not given any information about the corresponding species. We want to predict the most likely species for each of the measurements. In ML-speak, we say that we're given new *inputs* (here the length-width pairs) and that we want to predict the most likely *output* for each of the inputs.

We'll plot the new inputs in black for the moment.

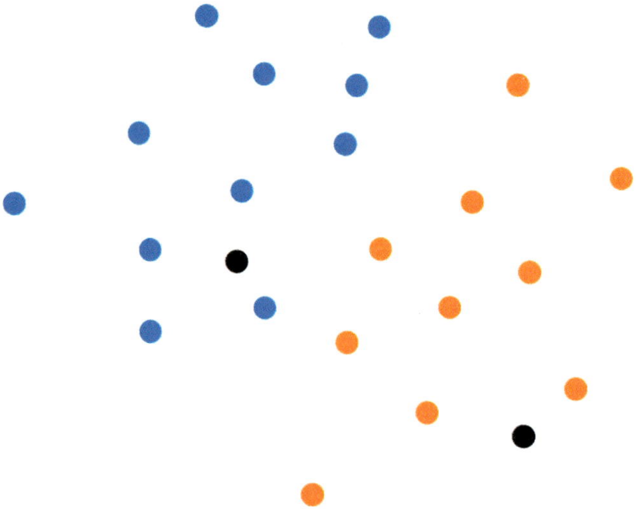

The color to assign to them, i.e., which predictions to make, looks fairly easy. The leftmost point should be blue and the rightmost point should be orange. So, let's just draw a line that separates the orange points from the blue points.

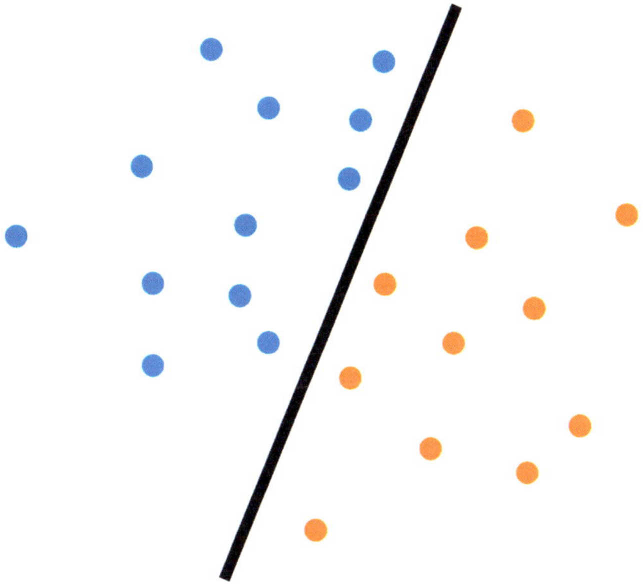

In the future we will affect colors based on which side of the black line the points fall on. That's it! That's our algorithm. What we're implicitly doing here is comparing new inputs to the example inputs we had, looking at the color (output) of those that are "in the same area." Our knowledge gathered from the examples that were given to us originally have taken the form of a line.

You will notice that there are several lines which would do the job of separating the orange examples from the blue examples. As you can see on the next figure, the line you choose to draw has an impact on some predictions, hence on the quality of predictions.

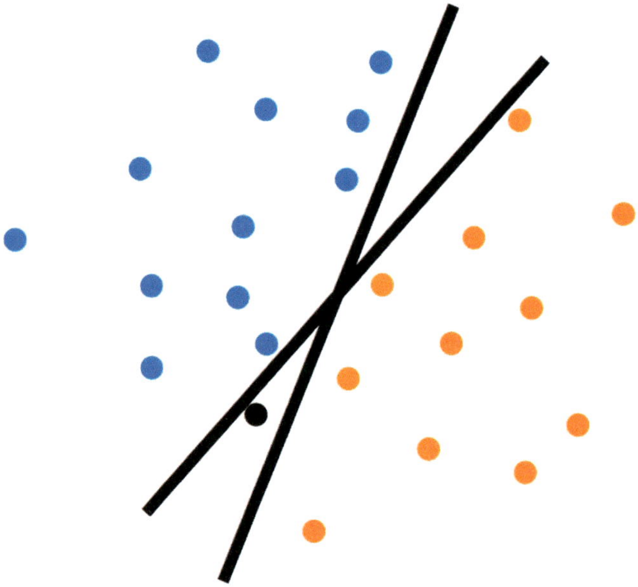

Machine Learning theory guides the choice of an optimal line. It would also allow us to find an optimal line of separation when the orange and blue populations, as plotted from the examples, overlap. And it can even find a non-linear boundary that would be better suited at separating the two populations.

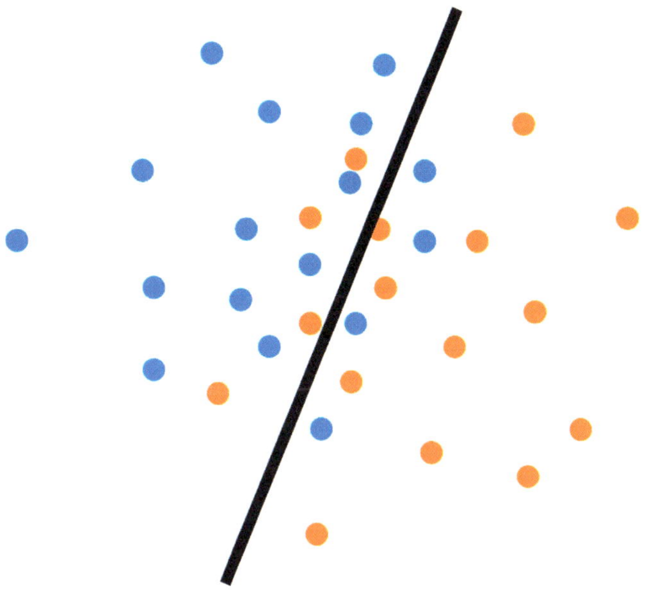

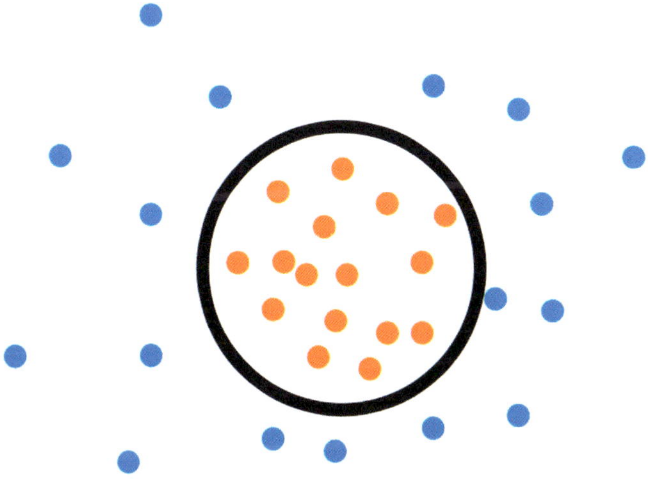

As you'll see later, this has been implemented for us in Prediction APIs, so we don't need to worry about how the boundary for separating orange from blue was actually found. We just need to know that it is determined from the examples.

LEARNING VS HAND-CODING THE RULES

Before Machine Learning came about, the way we categorized objects was with a set of hard-coded rules, or, a combination of if-then-else statements. For instance, we would have said," If the petal width is smaller than w1 and if the length is smaller than l1, then this is a setosa flower. Otherwise, if the width is smaller than w2," ... and so on.

This sort of thing has been done in the field of language processing, which consists of extracting meaningful information from text or spoken words. Algorithms were based on complex sets of expertly hand-crafted rules that required very deep insights into the problems at hand. There were lots of parameters to consider, and you had to think about all the possible cases.

The use of ML makes predictions much easier than the rule-based systems. ML algorithms are flexible since they create their own "rules" for the problem at hand, based on the analysis of example inputs and corresponding outputs. There is no need to code any rule by hand — just to provide examples. In the previous illustration of the functioning of an ML algorithm for categorizing objects, a simple rule was created by learning a boundary between the example inputs in the two categories: if a new input is on this side of the boundary, it is from one category, and if it is on the other side, it is from the other category. In the case of a linear boundary and for an input represented by a (x1, x2) pair, the rule takes the form:

```
if (a*x1 + b*x2 + c >= 0) then category1;
else category2;
```

This is only one rule, but a very "smart" one since it applies to a quantity (a*x1 + b*x2 + c) which is calculated from a combination of the input representation and constants (a, b and c) derived from an analysis of the examples.

The late 1980s saw a revolution in language processing with the introduction of ML that led to important improvements over rule-based systems. Some of the earliest ML algorithms, called "decision trees," automatically produced systems of hard if-then rules similar to existing hand-written rules. The more recent algorithms do not follow the structure of decision trees, which were mostly motivated by the desire to make predictions based on decisions that remained easily interpretable by humans.

Do you find yourself writing rules by hand? If so, it's time to try for a little ML!

As a side note, whereas we illustrated an ML algorithm that is based on a geometric argument, it is worth knowing that there is a whole other class of algorithms based on probabilistic arguments. I won't go into any details on their functioning since it would require notions of probability theory. These algorithms are popular in language processing because they are generally more robust when given unfamiliar or erroneous input, as is very common in the real world. They also produce more reliable results when integrated into larger systems comprising multiple subtasks, since they allow a fine modeling of the uncertainty in the predictions.

ML is flexible as it automatically creates its own rules for the problem at hand, based on an analysis of examples.

WHEN ML FAILS

It looks like, in theory, you can use ML to predict anything. You provide a few examples, let the algorithms figure out the rules, and then apply them to new inputs to get predictions. Hmm... I'm sure you didn't think it would be that easy, right?

REPRESENTATIVENESS OF TRAINING DATA

There are three notable things that can make ML fail. The first is when new objects aren't similar enough to any of the examples.

We say here that the machine fails to "generalize." In everyday language, generalizing means extending your experience of something to something else which is similar. As you saw in the previous categorization example, this is exactly what the machine does. The examples make up the machine's "experience," and the machine colors new points based on the color of close example points.

The generalization problem occurs when there are not enough examples, or when the examples are not representative enough. If the examples were representative, new inputs should never fall far from example inputs.

SIMILAR INPUTS, SIMILAR OUTPUTS?

Another thing that makes ML fail is when there are similar inputs that are associated to different outputs among the examples. When that happens in a classification problem, it becomes difficult to clearly separate classes.

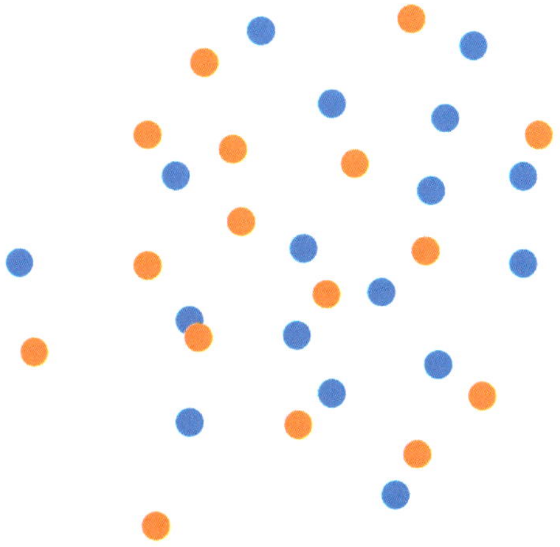

We previously assumed that there was a natural separation between classes. We started with a linear example and admitted we could deal with boundaries of any shape, but we still assumed that there was a boundary, which means that one class is in one region of the space and the other class is in another region. But why should things be this way?

When considering Machine Learning, we implicitly assume that there is a relationship between inputs and outputs. If this is indeed the case, then similar inputs should be associated to similar outputs and dissimilar inputs should be associated to dissimilar outputs. When visualizing the objects that we aim to classify in a space (for instance a 2D space in the previous illustrations), similarity is equivalent to proximity in that space. This explains why points that are of the same color should be relatively close to each other and far from points of the other color, thus leading to the existence of a boundary between the two colors.

The machine relies on that same assumption that similar inputs should be associated to the same outputs... at least, most of the time. It allows for some exceptions, but if these exceptions become the rule, the machine becomes unable to understand the underlying structure of the examples that were given to it. Typically, this happens when we haven't captured enough information in the input representation. We know that there is a relationship between the objects and the attribute we try to predict, but if the characterizations of these objects are too poor, the relationship won't be visible. We need to add more richer characterizations to our inputs so we can "break" the similarities of those who were associated to different outputs. In our previous example of flower categorization, this would be done by adding more measurements to petal length and width only, such as sepal length and width.

Initially, we only had 2 dimensions in this example (1 dimension is 1 attribute/aspect/characterization of the object you try to classify). This made it possible to visualize objects with a 2D plot. You can still visualize things when there are 3 dimensions, but what if there are more? It is common to have tens or hundreds of dimensions in ML problems. There is still a space associated to that, even though we can't visualize it. Objects of one class lie in one region of that space, and objects of the other class lie in the rest of the space. Maths can extend the notion of proximity to spaces of higher dimensions, determine boundaries, and tell you which side of the boundary an object is on by computing the sign of an expression which involves all of the object's N components (x1, x2, ... , xN). For a linear boundary, for instance, this expression has the following form:

- in 2D: a*x1 + b*x2 + c
- in 3D: a*x1 + b*x2 + c*x3 + d
- in 4D: a*x1 + b*x2 + c*x3 + d*x4 + e
- etc.

NOISE

This issue of having similar inputs associated to dissimilar outputs may persist for another reason: noise in the outputs. Noise is when the example outputs that we recorded are not what they should be. Let me illustrate this: we implicitly assumed that our flower examples had been correctly classified, but this had to be done by someone at some point, right? An expert botanist had to look at the example flowers and say, "This is a setosa, this is a virginica." The botanist thus acted as a "teacher" who made it possible for the machine to learn. But what if the teacher made mistakes? Besides, there is always a chance of introducing errors when recording data...

Noise is also found in measurements, which are always a bit off because our tools and sensors are imperfect and only work to a certain degree of precision. Other times, noise is caused by phenomena that are specific to the domain to which we want to apply predictions and that we fail to capture in the inputs. Ideally, the output should be completely determined from the input. But consider, for instance, that we record real estate transaction prices along with property attributes, in order to create examples for being able to predict the value of a property based on its attributes. There are circumstances that can bring a price below or above actual (true) market value, such as when a property needs to be sold quickly or when someone wants to buy a property that wasn't originally for sale. Algorithms are designed to deal with noise to a certain point, but as you can imagine, the more noise there is, the more random things look, and the harder it is to learn anything. In any case, the fact that we have "wrong" examples perturbs our ability to learn.

With these three cautionary remarks, it has become clear that our success in applying Machine Learning is tied to our ability to gather a good set of well-characterized examples...

CHAPTER 3: CONCEPTS

3

TYPES OF ML PROBLEMS

Learning consists of establishing relationships between example inputs and outputs. It is used to predict the output corresponding to a new input. In the flower categorization example in the previous chapter, inputs were arrays of two numbers, and outputs were labels. We could also consider inputs that are arrays which contain more numbers, or that contain both numbers and labels.

We saw that predictions are made by comparing a new input to example inputs. Inputs are compared by combining comparisons for each of their components. When components' values are numerical, we can compute differences. When they are labels, we can just say whether or not they are the same. We could even consider textual components and compare them by looking at the number of words or phrases in common between bits of text.

Our previous categorization problem was *binary* since there were only two possible categories. These are also called *classes*, and categorization is also called *classification*. One practical example of classification is tagging incoming emails or comments on a blog post as "spam" or "ham" (i.e., undesirable or regular). A class is identified by its label, but it could also be represented by a number. For instance, we could have -1 for one class and +1 for the other in a binary classification problem. We would thus be trying to predict an output which is a number equal to -1 or 1.

```
Spam prediction example
The input components are the email content (text)
and the sender's email address (label). The output
is either "spam" or "ham".

> Input: ["Lose weight now! Our new diet will
[…]", "magicdiet@betteryou.com"]
> Output: "spam"
```

```
Flower binary classification example
The input components are the width and length of
petals in centimeters. The output is either +1
(for iris virginica) or -1 (for iris setosa).

> Input: [10.3(cm), 4.5(cm)]
> Output: -1
```

Actually, why not try to predict numerical outputs that would be continuous and real-valued? Predicting a real-valued number is called *regression*. One example of regression is predicting house sale prices. Typically, the input is a list of house attributes such as the year the house was built, its surface, and the number of bedrooms. The output is the price of the house.

```
House sale prediction example
Input components are the year the house was built,
the surface in square meters, and the number of
bedrooms. The output is a value in dollars.

> Input: [1960; 103; 2]
> Output: 300000
```

In Natural Language Processing (NLP), inputs are purely textual. Outputs are often categorical — for instance, when we want to identify the topic of a document, the sentiment of a comment or of a review, and so forth.

```
Sentiment analysis example
The input is a piece of text. The output is either
"positive" or "negative".

> Input: "What a great piece of software"
> Output: "positive"
```

Outputs can also be structured, for example, when we want to extract a date from a sentence. But this would correspond to a different type of problem which won't be covered in this book (see Chapter 6 for pointers to APIs to help you tackle such problems as date and entity extraction).

THE TWO PILLARS OF ML

When we talk of a Machine Learning algorithm, we actually talk of two algorithms: a *training* algorithm and a *prediction* algorithm.

- Remember our illustration of the functioning of a classification algorithm? Training consists in learning a *model* from examples (*data*). A training algorithm (also called *learning algorithm*) takes a set of input-output examples, or *training dataset*, and creates a model by analyzing this dataset. In the previous illustration, the model is the boundary that separates the blue points from the orange points.

- A prediction algorithm takes a model and a new input, and gives back an output value. We make a prediction of an output for a given input *against* a model.

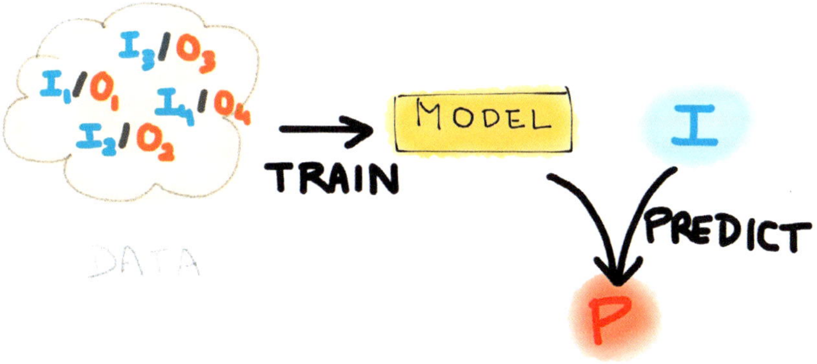

THE SPREADSHEET VIEW

Imagine that we import a set of data points which all have the same structure into a spreadsheet program. Each element of the dataset is an object that is stored in a row, and each column represents an aspect or attribute of the objects which provides information about them and allows to characterize them. The problems we can tackle with Machine Learning are predicting the values of missing attributes. These missing values are highlighted in orange and in pink in the following excerpt of a house's sale price spreadsheet.

Bedrooms	Bathrooms	Surface (foot²)	Year built	Type	Price ($)
3	1	860	1950	house	565,000
3	1	1012	1951	house	
2	1.5	968	1976	townhouse	447,000
4		1315	1950	house	648,000
3	2	1599	1964	house	
3	2	987	1951	townhouse	790,000
1	1	530	2007	condo	122,000
4	2	1574	1964	house	835,000
4			2001	house	855,000
3	2.5	1472	2005	house	
4	3.5	1714	2005	townhouse	
2	2	1113	1999	condo	
1		769	1999	condo	315,000

We can only pick one attribute to make predictions on at a time. In other words, we have to fix a column on which to make predictions (the last one in our example — price). We could fill any gap in any other column, but we would need a different model for each column. The attribute we want to predict becomes the output of our learning problem, and all the other attributes are the input components, usually called *features*. Each feature represents a piece of information that has an impact on the outputs to be predicted. The surface, number of bedrooms and year of construction are features that all have an impact on the price of a house.

The training dataset consists of all of the objects for which the output is known (highlighted in light yellow). The rest of the data is what we could call the new data and the cells highlighted in orange are the ones whose values we will be able to predict with the model built from the training data in white.

The spreadsheet is just a visualization of the data for your learning problem. Quite often, data is not fixed in a spreadsheet but comes in a stream (for instance when you observe users' behavior on your app). If you observe a new input-output pair, you add it to the training dataset, and you recompute a model based on this new training dataset. Algorithms usually do not recompute the predictive model from scratch, but instead, they adjust the previous model based on the new data: we say that we *update* the model with new data. Then, when a new input arrives, you use the updated model to predict the output.

ML can be used to **predict missing values in your data.**

THE SOFTWARE ENGINEER'S VIEW

Another way to see a Machine Learning system is as a program creating another program — more precisely, as a training program which creates a prediction program. What you care about ultimately is the program that makes predictions. Initially, before any data is fed to the system, the prediction program is just a "template" which relies on constants that haven't been defined yet. The training program uses data in order to set these constants. Let's look again at our illustration of the functioning of categorization/classification.

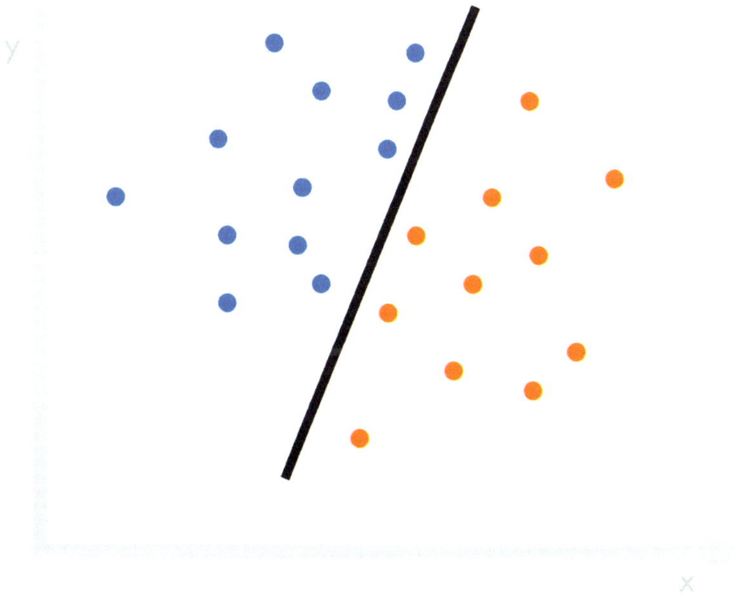

Predictions are made by looking on which side of a separating line an (x, y) input falls. Regardless of what the actual separating line is, the program for this has the following form:

```
if ( $1 * x + $2 * y + $3 >= 0 )
    return blue;
else
    return orange;
```

`$1`, `$2` and `$3` are placeholders for constants. These 4 lines of code form our template prediction program. The training phase consists of using the example data to determine a separating line. This means that the training program sets the values for `$1`, `$2` and `$3`. So, what was just a template has become an actual program that can be used to make predictions!

ML-AS-A-SERVICE

The easiest way to apply ML to some data is to a use black box program with training and prediction algorithms that have been implemented for you. "Black box" program means that you know what to give to the program, you know what to get from it, but you don't know how it works and what algorithms are being used. This is fine for testing and evaluating the accuracy of the predictions you could get (more on that in Chapter 5).

In production, however, you need to make sure you have adequate resources to run the training and prediction programs. What if the predictive model needs frequent updating because your application gathers a lot of training data, and what if you have a lot of predictions to make in your application because it's getting a lot of traffic from users? Will the black box programs running on your server handle the load?

If you don't have time to figure out how the ML algorithms work, chances are you don't have time to figure out how to scale them. Thus, the easiest way to apply ML is to use black box programs that run "in the cloud," so you don't need any infrastructure and scaling happens automagically. Let's call that *ML as a Service* (MLaaS).

TYPES OF APIS

MLaaS is accessed through APIs which come in three types:

- *Specialized Prediction APIs* realize specific predictive tasks such as detecting the language of a piece of text or detecting the sentiment of a Tweet (see list in the corresponding section of Chapter 6 for more examples). Specialized Prediction APIs are the easiest to use: you do not need to figure out what the best algorithm is for the task, you do not need to think about how to best characterize the objects to make predictions on, and you do not need to provide training data: all of this has been done for you. All you need to do is to give a raw representation of an object and you'll get a prediction back.

- *Generic Prediction APIs* allow you to tackle any classification or regression task. As you may remember from the beginning of this chapter, there are two phases in ML, one to learn a model based on training data and the other one to make predictions against that model on new data. This is reflected in Generic Prediction APIs because they include two core methods. The first one is used to send training data and get a model back, and the second one is used to ask for a prediction for a given input against a given model. You do not need to figure out what the best training and prediction algorithms are for the task at hand, but you do need to think about how to characterize the objects to make predictions on (i.e., which features to use), and you need to provide training data.

- *Algorithms APIs* give access to specific ML algorithms. This is like using a black-box algorithm, but in the cloud (so you don't run into scaling issues). In addition to what you have to do with a Generic Prediction API, here you need to tune the algorithm's parameters to the problem at hand, and you may also need to try a few different algorithms to see which work best.

RELATIONSHIPS BETWEEN THESE APIS

Algorithms APIs have the least abstraction, because both Generic and Specialized Prediction APIs could be based on them. They are easier to make but they are the hardest to use because you have to choose a training and a prediction algorithm for your problem and you need to tune their parameters.

It is easier to use Generic Prediction APIs because they analyze the training data to automatically figure out the best algorithms and the best parameters to use for the problem at hand. The technology behind this is relatively recent. The first API of this type was Google Prediction API, which was released to the public in 2011 (and only dealt with classification at that time). As you might imagine, it is difficult to create a meta-algorithm that can adapt to any problem you throw at it and that can select the type of algorithm/model that will work the best out of a number of possibilities. This difficulty shows in the number of available Generic Prediction APIs compared to other types of APIs.

Specialized Prediction APIs have the less versatility, but they are the easiest to use. They should be your first port of call when you want to make predictions that answer a specific question: did someone else already work on the same (or a similar) problem as you, and did they create a specialized API for it? If so, chances are that an expert spent time choosing the best algorithm based on his knowledge of the problem, the domain of application, and a lot of trial and error. This expert work still outperforms that of a

Generic Prediction API, so it's better (and quicker) to use Specialized Prediction APIs — unless you figure out later that your problem is not exactly the same as the original problem that the Specialized API tackles.

If you cannot find a Specialized API that meets your needs, you will have to use a Generic API and add training data to it. You can then create your own Specialized API if you want, based on the prediction method of the Generic API and on the model that it gave after training.

SNEAK PEAK

As you may have understood, this book focuses on teaching you how to use Generic Prediction APIs. I want to show you how simple it can be to create a model and to use it to make predictions, so I have included some actual code that uses the BigML API wrappers for Python:

```
from bigml.api import BigML

# create a model
api = BigML()
source = api.create_source('training_data.csv')
dataset = api.create_dataset(source)
model = api.create_model(dataset)

# make a prediction
prediction = api.create_prediction(model,
new_input)
print "Predicted output value:
",prediction['object']['output']
```

As you see, there are two parts in this code: one for training a model, and one for making predictions against this model.

Creating a model

`training_data.csv` is the CSV (Comma-Separated Values) file that contains the training data (no kidding!). It is formatted as follows: each row corresponds to a training data point and each column corresponds to a feature, except for the last column which corresponds to the output. By definition of a "training" data point, there cannot be missing values in the last column. It is common to use the first line of a CSV file for column headers. In the house price example, the first lines of `training_data.csv` would be:

```
beds,baths,surface,year,type,price
3,1,860,1950,"house",565000
2,1.5,968,1976,"townhouse",447000
4,2,1315,1950,"house",648000
3,2,987,1951,"townhouse",790000
1,1,530,2007,"condo",122000
```

This is the CSV version of the portion highlighted in yellow of the spreadsheet showed in the previous section. Each line corresponds to a property, and the columns are: number of bedrooms, number of bathrooms, surface in square feet, year of construction, type of house, and price in dollars (output).

Making a prediction

`new_input` needs to be a Python dictionary structure where keys correspond to headers, as they appear in the first row of `training_data.csv` (except for the last one which is for the output). For instance:

```
new_input = {"beds": 3, "baths": 3, "surface":
1503, "year": 2004, "type": "house"}
```

prediction is an object that contains various information, with the predicted output value for new_input being stored in prediction['object']['output']. This is a numerical value in the range we could expect when looking at the example outputs in training_data.csv; the output of this script could be:

```
Predicted output value:  677684.61538
```

You will find further information on BigML and other Prediction APIs in Chapter 6. Most of the code you'll write (if any) in addition to the previous lines will be to prepare training_data.csv — we will discuss how to create the best training data to have the best model for your own ML problem in Chapter 5. We will also discuss how to measure the quality of the predictions made with the model you created. But for now, let's review some example uses of Machine Learning.

CHAPTER 4: EXAMPLES

4

FORMAT

In the same way that we try to make machines learn from examples, I want you to learn what ML is about from example use cases, which is why I dedicated a full chapter to them. I chose to include it at this point of the book for two reasons:

- It is a great way to review the concepts and to practice the vocabulary we've introduced in the previous chapters.
- I want you to think about these examples when reading the rest of the book — more specifically, about how my advice would apply to them.

To tell you the truth, I am usually quite disappointed by the ML examples that I've read about. Most of the time they are too abstract. You never know which question the predictive model is exactly trying to answer, which are the inputs and outputs, and how the learning happens. Conversely, you can find very detailed examples in books on ML or Data Science, but they would be so detailed that it would take time to follow from beginning to end, or they would be intertwined with the rest of the content of the book. In this book, I have gathered examples in the same place so you can get an overview of ML's diverse uses and you can easily refer to them. In terms of manuscript length, I have tried to strike somewhere in the middle by giving quite a few examples, detailed over a couple of pages each, and structured with the following information:

- **Who**: who does the example concern?
- **Description**: what is the context, and what are we trying to do?
- **Question asked**: how would you write the questions that the predictive model should give answers to in plain English?
- **Type of ML problem**: classification or regression?
- **Input**: what are we doing predictions on?
- **Features**: which aspects of the inputs are we considering, and what kind of information do we have in their representation?
- **Output**: what does the predictive model return?
- **Data collection**: how are example input-output pairs obtained to train the predictive model?
- **How predictions are used**: when are predictions being made, and what do we do once we have them?

My intent through the examples I'm presenting in this chapter is to show you the potential of ML and the basis of what makes it work. Although they have been simplified for the sake of pedagogy and brevity, they will give you a better idea of what sorts of questions can be asked to predictive models, and you may be able to identify a problem similar to yours. I have divided these examples in two kinds: Business and Apps.

BUSINESS

In this section, I have grouped together examples where ML is used to improve a business by optimizing revenue or saving time. This first shows in the simple fact that, either in the inputs or outputs of these examples, there will be values that correspond to amounts of money! Also, the inputs will consist of customers and/or products.

ML creates models that make predictions. When using ML in apps, you would be using these predictions programmatically. In a business context, you may only be interested in having a descriptive predictive model but not interested in using predictions given by the model. The reason for this is that, whether data-driven or not, some decisions need to be made by actual people in the end. So, it makes sense for business decision-makers to use predictive models that can be interpreted by humans, just to gain understanding and new insights into the problem at hand.

Below is the list of examples in this section:

- Churn analysis
- Up-sell opportunity analysis
- Pricing optimization

- Sales optimization
- Fraud detection
- Credit scoring

CHURN ANALYSIS

- **Who**: Companies who sell subscriptions or monthly price plans, such as telecom companies.

- **Description**: We want to build a system that identifies customers who are at risk of leaving ("churn") within a given number of months X, so we can take immediate action to make them stay; higher customer retention will result in more revenue.

- **Question asked**: "Is this customer going to leave us within X months?"

- **Type of ML problem**: Binary classification.

- **Input**: Customer.

- **Features**: Individual "profile" based on information about the customer (e.g. age, income, house value, college education), characterization of his usage of the service (for the telecom example: average call duration, overcharges per month, leftover minutes per month), of his interactions with customer support (e.g. number of interactions, topics of questions asked, satisfaction ratings) and any other contextual info (e.g. handset price).

- **Output**: No-churn (negative) and churn (positive) classes.

- **Data collection**: Historical data made of all the customers we had X months ago, with feature values determined from the information we had then. Now, X months later, we know who has left so we know the associated output values.

- **How predictions are used**: Every X months, we build a new model or we add training data to the existing model (in which case, the same customer will be present several times in the data but as "snapshots" of him taken at different times). We apply the model to make predictions for each current customer (compared to last time, some may have left and some new ones may have joined). We give special offers to customers identified as positive by the system, to entice them to stay.

UP-SELL OPPORTUNITY ANALYSIS

- **Who**: Companies who sell a line of products.
- **Description**: If we've sold a product to a customer, chances are we can sell him more; we want to predict whether a customer is going to be interested in a new product, and take action if so (by sending a promotional email, for instance).
- **Question asked**: "Is this customer going to be interested in this product?"
- **Type of ML problem**: Classification or regression.
- **Input**: Customer or customer/product pair.
- **Features**: Info about customer (demographics), indicators of what else the customer bought in the past (one per product), info about product (e.g., category, description, price).
- **Output**: 1 if customer will show interest in product, 0 otherwise. Or, a label that characterizes this interest (e.g., "none", "product_page_view", "add_to_cart", "buy"), or a rating from 1 to 5, in which case we are doing regression.
- **Data collection**: Every time we have an opportunity to present a new product to a customer (either as they connect to the website or as they open an email sent to them) we see if they

show interest or not. The size of the training data is the number of customers who have seen the product.

- **How predictions are used**: We apply the model to make predictions for each current customer. We send promotional emails to customers identified as positive by the system (interested in the product) to tell them about the product, or we show a dedicated popover when they connect to our website.

PRICING OPTIMIZATION

- **Who**: Shops, stores, and sellers.

- **Description**: We are introducing a new product within an existing category of products that are already being sold, and we want to predict how we should price this new product; the product could be, for example, a bottle of wine in a wine shop, or a new house for sale.

- **Question asked**: "What should be the price of this new product, in this given (and fixed) category?"

- **Type of ML problem**: Regression.

- **Input**: Product.

- **Features**: Information about the product, specific to its category. In the wine bottle example, this could be the region or origin, the type of grapes, or the rating from a wine magazine. In the house example, this can be the number of bedrooms, bathrooms, the surface, the year it was built, or the type of house. We can also include a text description, and, when relevant, the cost to manufacture the product and the number of sales (total or per period of time).

- **Output**: Price.

- **Data collection**: Every time a product of the same category was sold, we log the price at which it went. Note that the same product might be sold several times (or not), and at the same or different prices, which affect the number of training data points.

- **How predictions are used**: We set the price of the product to the value given by the predictive model (no need to add a margin, this is already incorporated by the nature of the training data). Note that if the number of sales is one of the features, we need to do a manual estimation of this for the new product before we can make a prediction on it. Besides, since prices are likely to change over time, it is important to frequently update the predictive model with new data.

SALES OPTIMIZATION

- **Who**: Shops, stores, sellers.

- **Description**: We are introducing a new and non-unique product within an existing category of products that are already being sold, and we want to predict how the price will affect the number of sales. The product could be a bottle of wine in a wine shop.

- **Question asked**: "How many sales will we make with this product at this price, in this given (and fixed) category?"

- **Type of ML problem**: Regression.

- **Input**: (Product, price) pair.

- **Features**: See "Pricing Optimization" example for product features.

- **Output**: Number of sales.

- **Data collection**: There are as many data points as there are products being sold in the considered category. For each one, we extract the number of sales from sales records.

- **How predictions are used**: We try different values for the price, and plot how the number of sales and the revenue (equal to the price times the number of sales) change. We can then find the price that maximizes revenue. If we are looking for a product to resell among many possible products, we can try several product-price combinations, make predictions for each, and see which will be most profitable.

FRAUD DETECTION

- **Who**: Organizations who need to make payments when certain events happen, such as insurance companies when claims are filed.

- **Description**: Fraud yields unnecessary costs (payments are made whereas they shouldn't have), but because of the amount of activity going on (e.g., claims received), all cannot be analyzed manually, so we need to detect fraud automatically. Anomaly detection is similar to fraud detection in that we need to be able to detect rare events that can result in important damages, such as security attacks on web servers.

- **Question asked**: "Is this activity fraudulent?"

- **Type of ML problem**: Classification.

- **Input**: An "activity" (it can be a claim, a transaction, an event, etc.)

- **Features**: Characterization of the activity, usually including an amount of money and information about the people involved in the activity.

- **Output**: Yes (positive class) or no (negative);

- **Data collection**: Historical data consisting of all activities observed, most of which are legitimate, and some of which have been manually labelled as fraud after investigation. Because of the volume of activities observed, this hand-labeling is difficult to achieve and chances are that some activities weren't detected as fraudulent although they were (so there is noise in the data). Other ML techniques have been developed to circumvent this problem (such as clustering) but they are beyond the scope of this present book.

- **How predictions are used**: We make a prediction for every new activity and check that it is legitimate before making any payment in relation to this activity. If an activity is detected as fraudulent, we investigate it manually (with priority given by the amount of money associated to the activity and by the confidence of the prediction). Note that the behavior of fraudsters changes over time, so it is important to not rely completely on the predictions, and to keep a human eye on new activities and frequently update the model with new training data.

Note that insurance fraud saps more than $80 billion annually in the United States alone (Coalition Against Insurance Fraud, 2007), cutting across industries such as healthcare, auto insurance and workers compensation.

CREDIT SCORING

- **Who**: Banks and financing organizations who are in the business of lending money.

- **Description**: Because repayment defaults result in complications for the borrower and losses for the bank, they should be prevented as much as possible, and we want to be able to predict defaults based on loan/credit applications. In a similar vein, we could also look to predict the success of a Kickstarter project[1].

- **Question asked**: "Will this person fail to repay the credit he is applying for?"

- **Type of ML problem**: Classification.

- **Input**: Credit application.

- **Features**: Information about the borrower (age, property magnitude, job, revenue) and his banking situation (checking and savings status, balance, credit history); information about the credit (amount borrowed, duration, total amount to be paid); and information about purpose of the credit (type of project to be funded).

- **Output**: Yes (positive) or no (negative).

- **Data collection**: Historical data of all credits that were to be repaid by now or earlier, and output saying whether they were repaid or not.

- **How predictions are used**: Make a prediction for every new credit application and only give credit if we predict that it will be repaid.

APPS

Let's now look at how ML can be used within apps, whether mobile, web-based, for the desktop, or even for the car! ML enables new features that act as individual assistants for their users, and this make the apps "smarter." I'll first describe some commonalities across the examples that follow, in terms of the characteristics of their associated ML problems. The data often consists of observations of users' behavior, and is collected in a continuous manner. Models don't necessarily have to be updated in real-time, but predictions need to be delivered in a way that doesn't affect the app's response time: either predictions are precomputed or they are extra-fast. For a fixed number of users, the amount of collected data would increase with time. The number of users is also expected to grow, which makes the data grow even more, and makes the number of predictions per day grow too. As a consequence, it is necessary to use an ML system and an infrastructure that scale — this necessity makes Prediction APIs all the more interesting to an app developer.

Even though we have left the Business section of this chapter, you will see that there is always a business objective being served with the introduction of ML-based features in apps. In each example, I have emphasized what value is being delivered to the user, which implicitly creates value for the business that makes the app. There are a couple of other changes compared to the previous sections of this chapter. I have written longer examples, and I have given specific names of apps corresponding to the examples (due to their uniqueness) and I have included some visuals for you to get a better idea of how ML-based features integrate with apps' interfaces.

Here are the examples you'll find in this section:

- Priority inbox (important email detection).
- Car trip optimization.
- Tweet sentiment analysis.
- Crowd prediction.
- House value estimation.
- Language detection.

PRIORITY INBOX

- **Who**: Google Mail (and potentially any email provider).

- **Description**: Spam detection was the first application of ML to emails, which resulted in a clearer view of the inbox. It's only natural to look at more ways ML could help, so we'd like to automatically detect emails that are important to their recipient and mark them as such on the email client interface. For simplicity, we'll assume that we build a dedicated predictive model for a fixed user, and that we only have to make predictions for incoming emails that start a new thread. If an email belongs to an existing thread then its importance would be determined by the importance of the first email in that thread.

- **Question asked**: "Is this email important?"

- **Type of ML problem**: Classification.

- **Input**: Email.

- **Features**: Message features (subject, body, recipients) and sender features (email address, contact identification, ratio of emails sent to/received from this person, proportion of emails read out of those he sent previously, etc.)

- **Output**: "Important" or "regular."

- **Data collection**: Unlike spam, you can't really expect people to explicitly label all important emails as such when they get them. While it's difficult to know what makes an email important to a user prior to showing it to him, it is much easier to come up with a heuristic based on observations of what the user did with the email: did he ever open it, read it, reply to it, how long did it take him, etc. Such a heuristic can be used as a proxy to observing the user explicitly label important emails. Every new email the person receives is potentially a new data point, which is added to the training data when the user has logged in and has had a chance to interact with the email. For a single user, the size of the data is rather manageable, but for an actual system in production you would be multiplying by the number of users and could easily get into the millions of new data points per day.

- **How predictions are used**: A prediction is made for every incoming email that hasn't been caught by the spam detector. Again, this is reasonable for a single user, but it could turn into millions of predictions per day in production. The following illustration shows how emails marked as "Important" are shown on Gmail's interface. Users are also able to teach and give feedback to the predictive system, in case the prediction was incorrect, by marking emails as important or not.

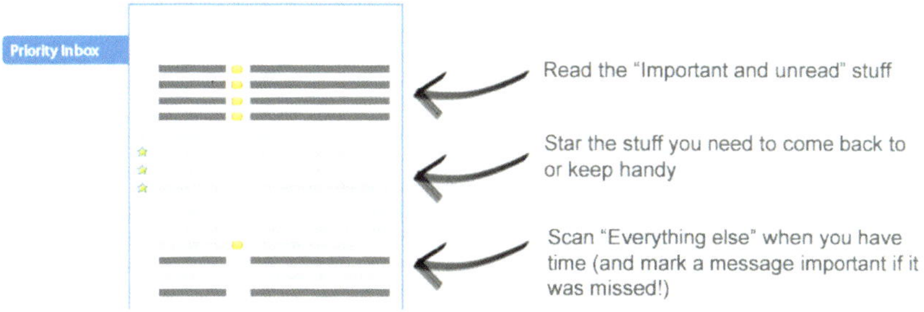

- **How value is created for the user**: Gmail claims that Priority Inbox saves you time[2]:

  "Looking at median time in conversation view, we noticed that typical Priority Inbox users spend 43% more time reading important mail compared to unimportant, and 15% less time reading email overall as compared to Gmail users who don't use Priority Inbox."

Sometimes, depending on your app, you will want people to spend more time on it rather than less. But in the case of email, the time savings introduced by an innovative feature are a strong argument to gain more users and to retain those who are thinking of leaving. Such features are essential for Google to stay ahead of the competition. Recently, they made Priority Inbox also classify emails into Social vs Promotions vs Updates vs Forums.

We will go over the Priority Inbox example in more detail in Chapter 7.

CAR TRIP OPTIMIZATION

- **Who**: Automakers such as Ford, who built a prototype in 2011.
- **Description**: The use of technology in cars either makes it easier for people to go where they need to go, or optimizes energy efficiency. Having an app that runs in the car and predicts where its driver is heading to would lead to progress in both areas, because the car could optimize itself for the trip ahead (think of the energy consumption parameters of hybrid vehicles) and assist the driver along the way.
- **Question asked**: "Where is this driver headed to? Is it one of his usual destinations?"
- **Type of ML problem**: Classification.
- **Input**: Context in which the driver starts the vehicle.
- **Features**: Origin (i.e., current location), and a list of X previous destinations and when they occurred. The origin and destination are location labels (e.g., "home", "store", "work", "parents") that have been explicitly defined by the driver or created by an algorithm that analyzed the car's driving history.
- **Output**: The destination as a location label.
- **Data collection**: We get one new data point after each trip. Some preprocessing needs to be done to create feature values.

We need to convert positions (given by a GPS as a latitude and a longitude) into location tags. This is necessary to keep learning the driver's personal habits (he may move houses or change jobs for instance) and to regularly update the predictive model with new data.

- **How predictions are used**: A prediction is made every time the driver unlocks the car or starts the engine. The system immediately computes a route to the predicted destination. When the driver is ready to go, the system asks him whether his prediction is correct (e.g., "Are you driving home?") and the driver can confirm or rectify. If the prediction is correct, the route that has been precomputed is shown. Otherwise, a list of top alternative destinations is displayed on the dashboard's screen. Once the car knows the destination, it can adapt its parameters to the trip ahead.

- **How value is created for the user**: The driver is given an estimate of the trip's time and a route that is optimized to the current traffic conditions, all without him having to give any input. Hybrid cars can also adapt fuel and electricity consumption to the route, and thus cost less money to their owners in the long term. Also, it's quite plausible that in the near future, cities will have Electrical Vehicle driving zones, so it would be useful for hybrid cars to adapt by using more fuel and recharging their battery prior to entering EV zones.

In 2013, Apple and Google introduced a feature on their smartphones that shows users the time to their predicted destination.

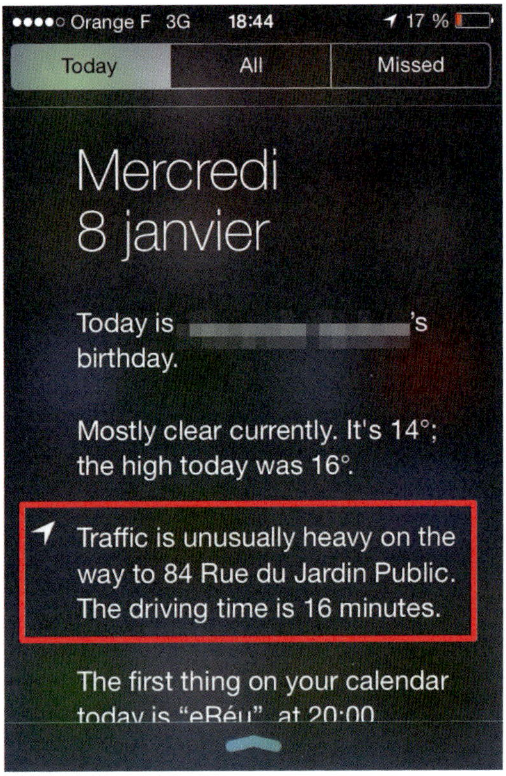

TWEET SENTIMENT ANALYSIS

- **Who**: mention.net (or any other social media monitoring tools).

- **Description**: Twitter users collectively send over 400 million tweets per day, a good portion of which mention brands. Consumers turn to Twitter to voice public complaints about brands (and expect that they will be listening). On the other hand, social networks are critical components in marketing campaigns and are used to relay positive messages about a brand. When managing a brand, it is thus essential to monitor all of this activity. So, we need to be able to recognize the human emotions expressed in tweets that mention our brand. Due to the high volume of tweets that must be monitored every day and their instantaneous nature, speed is of particular importance, so we turn to Machine Learning to analyze sentiment.

- **Question asked**: "What is the sentiment of this tweet?"

- **Type of ML problem**: Classification.

- **Input**: Tweet.

- **Features**: Text of the tweet, pre-processed (Tweets usually contain slang, emoticons, @ mentions, # hashtags and twit-jargon); punctuation (it helps convey sentiment and thus should not be discarded as it is in other text analysis tasks); number of other tweets in the discussion, and how many positive/neutral/negative; indicator of the presence of a link and type of page/site linked to (support forum, brand's website, blog, etc.); and an indicator of the presence of other people mentioned in the tweet.

- **Output**: "Positive", "neutral" or "negative."

- **Data collection**: Tweets need to be collected from the Twitter API and then manually labelled as positive/neutral/negative. When doing so, it is preferable to filter out tweets that are ambiguous or that contain mixed sentiments, and to balance the dataset so that each class is equally represented, in order to not skew the model towards a particular class. It is common to build predictive models for sentiment analysis from several tens of thousands of examples, and labeling that many examples is a long and costly process. Fortunately, sentiment analysis is a fairly standard problem for which there are specialized APIs where the training has been done for you (see Chapter 6).

- **How predictions are used**: A social media monitoring tool first gathers Tweets in which our brand is mentioned, and we then make a prediction for each one. This should be done in real-time so that we can take rapid action after detecting a negative tweet (and thus transform negative publicity into an opportunity for positive press), or we can capitalize on an opportunity after detecting a positive tweet. Being able to measure the proportion of positive Tweets also helps to measure the success of marketing campaigns and benchmark our brand. The number of Tweets to analyze depends on how popular our brand is, but the more popular it is, the more likely we are to have resources to do this analysis!

- Remember that any text documents can be subject to sentiment analysis. For instance, you could use predictive models within your company to gauge employee satisfaction, predict attrition, or to monitor the health of a project, provided you have enough resources to manually label example documents.

CROWD PREDICTION

- **Who**: SNCF, the French rail services operator. This example also extends to public transport services, car-park, bike scheme, city operators, and any operator who has fixed resources (trains, bikes, or space) and faces fluctuating demand.

- **Description**: There is already a web and mobile app made by SNCF that gives access to real-time information on train statuses and departure boards, and lets users search for travel routes/schedules. We can gauge demand in real-time through search queries. Thanks to the mobile app, we can also estimate how busy train stations and services are, based on position tracking (if allowed by users), check-ins at train stations, or explicit feedback such as "the train is full". We want to go one step further than real-time and give users predictions of how busy a future train is going to be, so that they can plan ahead and ultimately the peak hours can be spread out.

- **Question asked**: "How many people will be boarding/alighting this given train at this station at this date and time?"

- **Type of ML problem**: Regression (two problems: one for boarding and one for alighting).

- **Input**: Train, station, day and month, and time.

- **Features**: Information about the train (an id + a line id + a direction), the station, the day of the week, a holiday indicator, the number of search queries in which this train (stopping at this station on this day at this time) was returned, the number of such queries where the station was the start/end point of the journey, and the number of checkins at this station on this day at around this time. Snips, the company behind SNCF's app[3], is also adding to that contextual information based on Open Street Maps, GTFS schedules[4], weather, socio-demographic and economic data.

- **Output**: Number of people who board/alight.

- **Data collection**: Historical data collected by the operator (passenger counts) and by the app (check-ins, searches, feedback).

- **How predictions are used**: Predictions are queried through a dedicated SNCF mobile app (Tranquilien[5]): the user starts by searching for a train schedule, and he is then presented a list of 10 trains departing around the specified time, and a prediction is made for each stop of each train (color-

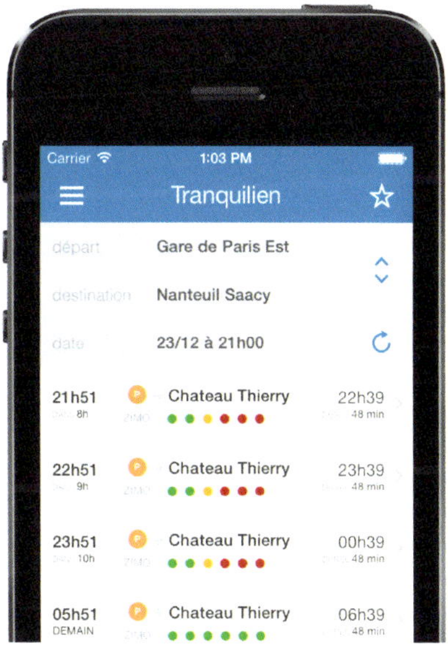

coded in green, orange and red). On average, a user does 2 queries per day, each returning 10 trains with about 10 stops, which makes about 2*10*10=200 predictions per day per user. SNCF aims to reach 300,000 users in 2014.

- **How value is created for the user**: Predictions allow commuters to better plan their journeys by seeing where they can easily find a seat on each train. This results in less frustration, less stress, and an overall better experience with the service. Being able to diffuse the load is also beneficial for the business as it results in better planning and service operation.

This example fits within a domain of application of new technologies which is simply called "smart cities."

HOUSE VALUE ESTIMATION

- **Who**: Zillow[6], Redfin[7], and other real-estate websites.

- **Description**: Real-estate websites are used by two kinds of people: buyers and sellers. Buyers need to know that the asking price of a property for sale is justified, given its characteristics and in comparison with other properties on the market. Sellers need to find the asking price that will maximize their chances of making a good sale. Therefore, it is beneficial to estimate the value of real-estate properties based on data on previous transactions, and to show estimates to buyers and sellers.

- **Question asked**: "At which price would a property with these characteristics sell?"

- **Type of ML problem**: Regression.

- **Input**: Real estate property.

- **Features**: Number of bedrooms and bathrooms, surface, year built, type of property (house, townhouse, condo), ZIP code, proximity to good schools, public transports, and access to shops and amenities.

- **Output**: Price (in dollars).

- **Data collection**: Information about real estate properties is centralized on Multiple Listing Services (MLS) and includes prices from previous transactions, which are used to provide data to train on. Because the market evolves, it is important to keep adding new examples of recent transactions to the training data and to discard examples that have become too old.

- **How predictions are used**: Every time a new property appears on the market, a new webpage is created and an estimate of the price/value of the property is computed. This estimate is shown alongside other information about the property, and the actual price at which it is announced. This price estimate is called the "zestimate" on Zillow. If a seller is advertising his property for sale, he can also ask for an estimate of its value. The estimate would get more accurate as the seller enters more information about the property, and this could drive him to provide more info. If a property has been on sale for too long, we can ask our predictive model for a new estimate of the value of the property based on the latest training dataset.

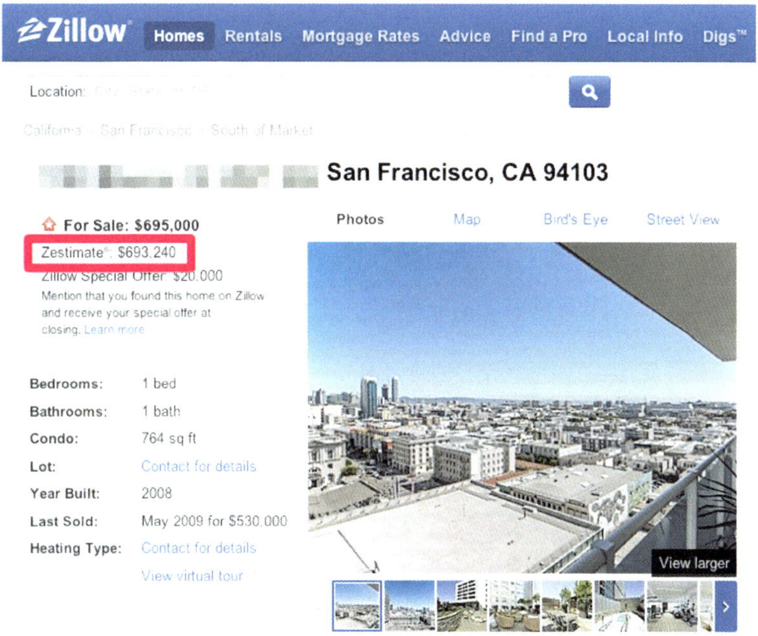
Zillow
Homes
Rentals
Mortgage Rates
Advice
Find a Pro
Local Info
Digs™
Location:
San Francisco, CA 94103
For Sale: $695,000
Zestimate®: $693,240
Zillow Special Offer: $20,000
Mention that you found this home on Zillow and receive your special offer at closing. Learn more
Bedrooms: 1 bed
Bathrooms: 1 bath
Condo: 764 sq ft
Lot: Contact for details
Year Built: 2008
Last Sold: May 2009 for $530,000
Heating Type: Contact for details
View virtual tour
Photos
Map
Bird's Eye
Street View
View larger

LANGUAGE DETECTION

- **Who**: Google Translate, or any text analysis tool or service that requires knowing the text's language before functioning — for instance, a sentiment analysis tool.

- **Description**: Google Translate is a translation service where a user can enter a piece of text in a web form, choose a source language and a target language, and get a translation of the text into the target language. Sometimes, however, users may not know the source language because they may browse the web, follow a few links, and get onto a website in a foreign language that they do not know. In these cases, Google Translate is used to tell the user the language he is looking at.

- **Question asked**: "In what language is this piece of text?"

- **Input**: Text.

- **Features**: Just text. The prediction service will use its own comparison method or feature extraction method (see Chapter 5).

- **Output**: One language out of a fixed set of languages like French, English, or Spanish.

- **Data collection**: We could constitute a set of examples of text-language associations by simply logging the queries to the translation service, where people would enter pieces of text and select the source language. These examples would then be used to train a predictive model. The queries would usually contain short pieces of text, ranging from a single word to a few sentences.

- **How predictions are used**: Our predictive model can be used in our translation service to detect the language of a piece of text entered by the user and automatically position the selection of the source language drop-down menu to the language that has been detected. Also, when the user browses the Internet with Google's browser, Chrome, we can have the predictive model run for each new page that is loaded, and we can prompt the user for a translation if the detected language is different from his own language. It is important to remember that the training data consisted of short pieces of text, so the predictions will work best on pieces of text of similar length. As a consequence, we can break a long piece of text into several shorter ones, make a prediction for each, and take a majority vote of the predictions. In case the prediction is incorrect, there should be a simple way for the user to let us know (see the screenshot that follows) which would result in new training data to use to update the model. Finally, note that predictions will only "work" for languages for which we have training data.

- **Other uses**: Imagine that customers write to your multinational brand at contact@yourbigbrand.com. You would be receiving emails in several different languages, which you would automatically detect in order to route each email to a customer service person who has the appropriate language skills.

CHAPTER 5: APPLYING MACHINE LEARNING TO YOUR DOMAIN

5

SPECIFYING THE RIGHT PROBLEM

FORMULATE A QUESTION TO PREDICT ANSWERS TO

Let's do a quick recap of the previous chapters. You can think of predictions as a way to answer questions about objects: "Is this email spam or not?", "How much is this house worth?", "What is the missing value in this row (in a given spreadsheet)?", etc. To predict the answers to these questions, we use Machine Learning algorithms and give them examples of objects and the question's answer for each example. Predictions are made on new objects by comparing them to example objects. Thanks to MLaaS, you do not need to know how the algorithms work, but you do need to know precisely which question you want to ask. This is done through the *predict* method of a Generic Prediction API or through a Specialized Prediction API.

The types of questions that we are interested in are those that are not obvious to find answers to. These are the types of questions you would ask to a crystal ball — except that you'd expect a crystal ball to "understand" what you mean, but here, you'll need to be very precise. You should formulate a question whose answer is either a number or a category/class/label. The examples in Chapter 4 should provide some inspiration. Try to formulate a question similar to one of these examples. If the question is categorical, your ML problem will be a classification problem and you should ask yourself what the possible categories/classes are. If the question is numerical, your ML problem will be a regression problem and you should ask yourself what the range of possible output values is.

COLLECT TRAINING DATA

ML needs training data to work. When you have an idea for an ML-based predictive system, you must find a way to collect data to learn from. This is kind of like finding a way to reach customers when you have an idea for a new product/business: without it, the idea falls apart.

For anything that involves observing how users interact with an app, data is collected in a natural and continuous way as people use the app. In all cases, we need to gather example inputs along with their associated outputs, which can be an expensive process. Consider, for instance, that we want to build a system that automatically figures out the topic of documents (such as business, political, and sports news articles). Getting example documents is relatively easy, but you then need to know the topic of each one, which can only be determined manually at first.

One way to get your example outputs is to hire people to label your example inputs through a service like Amazon Mechanical Turk[8]. But before doing that, I would recommend looking at whether people have worked on a similar problem in the past and gathered data you could reuse (whether for free or a fee). Datasets are subject to copyright law, so if you download a dataset for free and are in doubt, contact the authors for permission to reuse. If the data you need to collect is in webpages, check out Import.io. If you want to have access to public data, try Enigma.io. Otherwise, you could search the Microsoft Azure Data Marketplace[9].

If the labeling of example inputs is done by different people, you should make sure that labels are assigned consistently. One example where this could fail would be when collecting training data for a predictive system that would detect the "spamminess" of an email with an index between 1 and 10. Besides the fact that it would take longer to label example emails than if you just considered spam vs ham, you can be pretty sure that different people will have different notions of the degree of spamminess of a given email.

MAKING PREDICTIONS... SO WHAT?

> *"When it comes to data, I like to think of the golden question: 'What action will I take once I have this data?' If the answer is no action, it's generally a bad idea to even consider looking at the data."* — Hiten Shah

> *"Gathering data is important, but it's crucial to be able to differentiate between data that's merely interesting and data that actually has leverage."* — Bruce Ernst

Once you have been able to formulate a precise question to predict answers for, and you've made a way to collect data to learn from, you'll get an idea of your predictive system's feasibility. Remember, not being able to get the example inputs AND outputs to learn from is the first reason why an idea for a predictive system may not work. You would then further test your idea by working on the best representation of objects on which to make predictions ("feature engineering") by preparing the data to send to a Prediction API, and then finally measuring the quality of your results. All of this will be covered in the rest of this chapter.

But before doing any of that, it's essential to make sure you're asking the right question. The only way to do this is to step back for a moment and reflect on what would happen next if your predictive system worked perfectly. Think about the value you could get from predictions, and how you would get it. How, exactly, will you use the predictions? How will they affect the rest of what you're doing? Which objective will they serve? Maybe what you would do with these predictions depends on their accuracy — what if the accuracy is not as high as you hoped?

In cases such as the house price prediction and the language detection examples, your objective would simply be to make accurate predictions (we'll see how to measure this in the last section of this Chapter). But most often, the accuracy of predictions is not an objective in itself. In Machine Learning that Matters[10], Dr Kiri Wagstaff calls for metrics that reflect the real-world impact of predictive models, and you, too, should measure the impact on your app's or your business's objectives. You should look at the bigger picture and measure that ML is doing something useful.

Predicting churn... so what?

Let's illustrate this idea with the churn detection example we introduced in the previous chapter. One of your goals as a business is to maximize revenue. For that, you could either invest your money in campaigns to acquire more customers, or you could do something to retain customers who are at risk of leaving. All the efforts you make towards customer retention have a cost, and you are interested in measuring the Return On Investment (ROI), which is the ratio between the extra revenue that results from these efforts and their cost. Customer retention is difficult because the company usually has a lot of customers and cannot afford to spend much time on each and every one of them. Otherwise, the costs would be too high and would outweigh the extra revenue overall.

Building a predictive model that would detect customers at risk of leaving could help reduce customer retention efforts by directing them towards these customers only, but it would just be one half of the solution. The other half is the action you take in order to make these customers stay. You could either contact customers to find out more about their needs and switch them to a different subscription plan that would work better, or you could give them a special offer or a promotion. But which offer would that be, and how much would it cost? And how often would the actions taken to prevent churn work?

Let's make some assumptions in order to illustrate how the answers to these questions would impact the Return On Investment of our ML-backed customer retention program. Let's say that we give special offers to customers detected by the predictive model, that the success rate of the special offer in preventing the customers to leave is 50%, that the cost of giving a special offer is $2, and that the customer lifetime value (the revenue attributed to the entire future relationship with a customer) is estimated at $100 on average.

We need to introduce some terminology to the predictions made by the model. A "Positive" is a customer who the model thinks is likely to leave, a "True Positive" (TP) is one such customer who will indeed leave, and a "False Positive" (FP) is one who won't actually leave. We can now break down the extra revenue we expect with our strategy, in comparison to doing nothing to prevent churn, as follows:

- for each TP: $100 * 50% - $2 = $48;
- for each FP: - $2 (you wasted money giving that special offer to someone who wouldn't have churned anyway);
- for each Negative: $0.

FP

(False Positive)

The revenue is then calculated by multiplying these amounts to the counts of True/False Positives/Negatives and by summing. You divide that by the cost of the churn program (special offers given + overheads) and you have the ROI, which you can compare to the ROI of customer acquisition efforts. It should be greater than 1, or else it's making you lose money.

It's interesting to consider the extreme case where you would have a perfect churn detection model but a very low success rate in preventing the detected customers to leave — say, 2%. Because the model is perfect, there would be no FPs, and for each TP the extra revenue would be $100 * 2% - $2 = $0, which would give an ROI of 0 and would render ML completely useless here. This example highlights the importance of what you plan to do (and what you can do) with predictions.

Decide on performance criteria first

You should frame predictive modeling projects around the specific value they could create. For a business, the value usually has to do with time saved or additional revenue. For an app, it has to do with increasing usage, either from existing users or new users joining. As you see, a discussion needs to happen at the business level in your company in order to decide how the predictive model will be integrated, to set specific objectives, and to decide on criteria that will be used to assess the performance of the model. If the answers to the previous questions are not clear ("How will you use the predictions? How will they affect what you're doing? Which objective will they serve?"), you may be onto something that's cool, but maybe not that useful — it's better to realize that now than after spending hours building a predictive model. Predictive modeling projects are iterative, so it may take a few attempts before you get somewhere. It helps to have a long term goal in mind and a clear value proposition.

We will discuss performance measurement again towards the end of this chapter, and among other things, we will see how to estimate performance in advance of a deployment into production.

Figure out which question to ask, how to collect data, **what value predictions will create, and how.** *Then build a predictive model.*

FEATURE ENGINEERING

FINDING THE RIGHT FEATURES

Remember what I said earlier about having a crystal ball that could make predictions? Try to do the same as that crystal ball yourself. How would you go about it? You don't need to know precisely: just saying, "I would consider *this* and *that*," is enough. *This* and *that* are pieces of information that you already have about the objects on which to make predictions, or information that you could derive from what you already have, and they should have an influence on the output. *This* and *that* are going to be your input components that characterize objects. They are going to be your *features*. As we said in Chapter 2, ML can automatically create rules to make predictions. You don't need to create these rules yourself, but you do need to extract a set of feature values to represent objects and that the rules will use. Figuring out which aspects of objects have an influence on the output requires domain knowledge and insights into the problem at hand, which is why you — not an ML expert — are the best person for this job. The activity of finding "good" features is called *feature engineering*.

It is rare that the features we want to use are all in the original objects' representation. We often need to do some preprocessing, where we take raw data from a source (a spreadsheet file or a database) and prepare a dataset of feature values and output values that can be used to train a model, i.e., a `training_data.csv` file. For predicting the price of a house, you would consider its surface, its location (latitude and longitude), the year it was built, and the number of rooms. You could also consider the proximity to public transports. This information has to be derived from a database of locations of public transport stations and from the location of the house. This database is an external source of data that you can use to enrich your own. For detecting important emails coming to your inbox, you would consider the email text, the subject, who the sender is, but also how many times you have written to the sender before, which again requires some processing of your email data. In Computer Vision (CV), the objects you make predictions on are images. If you would use raw images you would be working on matrices of pixels, or on tables of cells with numerical values in them. Instead, you need to work with features that are invariant to basic transformations of images such as scale and rotation, because in most CV problems, they won't make the output change.

For clarity, try to make your features as "independent" from one another as possible. This way, the value of one feature cannot be deduced from 2 other features. When it makes sense, numerical features should be *normalized*, meaning they should be scaled to range between 0 and 1. This makes it easier for the system to identify input-output relationships and create rules for prediction. Let's go back to our important email detection example. My first intuition is that important people tend to send important emails. If, among all my conversations with a given contact, more were started by me than by him, then this contact is likely to be important. Instead of using the number of conversations started by me and the number started by him as features, I'll use only one feature, which is the proportion of conversations that were started by me (between 0 and 1). Does this mean you should always normalize numerical values? No: if your numerical value is "absolute", like the year a house was built, normalizing won't help and will only make it harder for humans to read the data.

Now, remember one of the reasons why ML fails: not enough good features. Extracting features is your first most important job. They must describe inputs in such a way that it is reasonable to assume that the output is (almost) completely determined from the input. Again, before you ask a machine to understand your data and make predictions from it, you should try that yourself. With the features you have, would you be able to make the predictions yourself? If you're not sure whether a particular feature is going to be useful or not, don't fret too much, because it will be discarded easily as ML algorithms are good at ignoring useless features. The more features, the better.

But simply breaking existing features into more features won't improve predictions. For instance, if you have a set of labels to describe an object, you would have one "labels" textual feature that would be a concatenation of the object's labels. However, if you instead use one "label indicator" feature per label, you would have as many features as the total number of possible labels, but it wouldn't improve the characterization of objects and thus wouldn't improve the performance of the system.

Extracting features is your first most important job.

You — not an ML expert — are the best person for it.

TYPES OF FEATURES

At first, you shouldn't hamper your creativity worrying about how features will be represented. Just think about the aspects that are important to consider in order to make predictions. I know a startup who crawls the web and automatically detects product pages of e-commerce websites. I don't know for sure how they do it, but my first guess would be to consider the structure of the webpage, the main image of the page, and the text content in order to predict whether it is a product page or not. Then I can start thinking about how to represent these features. The structure of the webpage would be the DOM tree, the image would be a matrix of pixels, and the text content would be made of a concatenation of the title of the page, the headings, and the paragraphs.

Here comes the bad news: most Generic Prediction APIs don't understand trees, nor sets, nor arrays of variable length. The types of features they work with are:

- Numbers (e.g., 1, 2, 3.0, 3.1, etc.)
- Labels (e.g. "low", "middle", "high")
- Text (e.g. "this string is not a label but a sequence of words")

You can work with pixel matrices since they are just arrays of fixed length, so each pixel would be a feature, but the API won't "see" the image that all these pixels represent (contours, shapes, objects, etc.)

For now, we would have to think of other features for our product webpage detection task. In the future, we may be able to use our original feature ideas if we decide to go beyond the simple use of Prediction APIs and to hire an ML expert. An expert will be able to devise a way to use DOM trees within appropriate ML algorithms and extract the visual features that matter from the images. The resulting predictive system will probably detect product webpages more accurately, and the first attempt with Prediction APIs will provide a baseline for comparison.

Sometimes, it takes some very simple processing to convert your features to types which can be used in Generic Prediction APIs, such as:

- A date can be represented by a day counter since a particular date of reference (that would be an integer), or by four separate features. These would be the day of week ("Monday", "Tuesday", etc.), day of month (integer between 1 and 31), month (either a class, e.g., "January", or an integer, e.g. 1), and year (integer).

- Time can be represented by a single decimal number between 0 and 24, or by three different features: hour, minute, and second.

- A set or an array of variable length can be transformed to a concatenation of string representations of its elements.

PREPARING THE DATA

Making sure you have good quality (training) data that follows the structure set by your choice of features is your second most important job. It is second because it follows the first job of feature engineering, but it is the most critical and complex job, so it's where you'll be spending most of your time. People in the ML community often say, "Garbage In, Garbage Out (GIGO)." This means that an ML system is only as good as the data it was given to train on. If you recall the example code given in the MLaaS section of Chapter 3, the model is completely determined by the `training_data.csv` which is sent to the API.

Preparing the data consists of three steps. The first step is to collect it and extract features. Sometimes it will only take a join of relational databases to get the desired dataset, and other times you will need to run a script through existing data to compute the desired features. The second step is to run sanity checks on the extracted dataset and to clean it up if necessary. The third step is to make sure the data is representative and thus ready to be used for training a model and estimating its future performance on new data.

HOW MUCH DATA IS NECESSARY?

A rule of thumb is to have at least 10 times as many examples as features. Good training data typically includes several hundred examples at least, and it can go up to millions of examples in big systems. For each categorical feature you have, make sure that you have at least 10 examples per category — ideally, you should aim to have a few hundred if you want really good predictions.

One way to achieve that is to merge related categories. Imagine that you want to predict the interest of individuals in a consumer product, and that, among other things, you want to take into account where they live. You could have a "country" feature, but you could also consider broader regions of the world, like Western/Central/Eastern Europe, North/South America, and North/Sub-Saharan Africa. If you think it is important to know the actual country, then make sure you have enough examples per country!

For a classification problem, you should also make sure that you have at least 10 examples per class.

WORKING WITH TEXT

If you are considering textual features, there are likely ways to transform the text values that will improve the quality of predictions.

ML works by comparing new inputs to example inputs. The comparison of textual feature values is based on how often the same words are to be found. Think of a word as "anything in between two separators", where a separator is a whitespace or the beginning or end of a piece of text. This has two consequences.

1. "word." is not matched to "word", so you should remove punctuation (other than apostrophes).

2. In proper names (or anywhere you would want a string to remain unsplit) you should replace white spaces by underscores (e.g., New_York).

Some algorithms actually look at how often the same groups of words of size n (also called *n-grams*, where n is a parameter of the training algorithm) are to be found. By default, n is equal to 1. In that case, you can represent sets of labels by strings in which the labels are concatenated in any order (e.g., "comedy animation black_and_white" for a movie).

Comparisons are case sensitive, so you should remove capitals at the beginning of sentences. For certain tasks, you may realize that stop words are useless (e.g., "of", "the", "at", etc.), so you should remove them. Or it may be that only the stems of words matter (e.g., "great" instead of "greatness", "fish" instead of "fishing", etc.), in which case you should transform the text to only keep the stems.

Finally, because training data should be representative, the size of the text feature values in your prediction queries should be about the same as those in your training data.

SANITY CHECKS AND CLEAN UP

> *"Data are just summaries of thousands of stories — tell a few of those stories to help make the data meaningful."* — Chip and Dan Heath

Manual checks

The idea here is to make sure that your data is as you think it should be and that there weren't any bugs in the data collection and feature extraction processes. The first things to check are:

- The type of values (e.g., numbers are indeed numbers — look out for NaNs).
- The range of numerical values (for features, and also outputs in regression problems).
- The distribution of input and output values (with a histogram or with a pie chart for categorical values).

Histograms automatically generated by BigML after a dataset has been created/imported.

Then, list all the domain-specific assumptions you can make about the data — even the most basic ones. The easiest and quickest way to check whether these assumptions hold is by browsing the data in Excel (or any other spreadsheet program) and sorting and filtering the column values. Excel can be slow when working with spreadsheets of a few MB. In that case, take a small percentage of the original data and work on that instead. You can either take the first lines of the data, or you can sample randomly, depending on what makes most sense in your case.

If you see data points that surprise you (maybe because a feature value is very high), have a look at them individually and try to understand why you have them in your data. You should also check whether there are any missing values. If so, can you fill them in, and do you understand why they are missing? While you will find ML algorithms and Prediction APIs (such as

BigML) that have no problem with empty/missing feature values, others (such as Google Prediction API) will evaluate the empty value as the empty string ("") for text and labels, and as 0 for numerical features. There will be no way to differentiate between an unknown value and a "" or a 0 in the data. These values would likely be different from what the actual feature values should be, which would perturb the learning of a model.

Scripts

It's a good idea to browse your data in a spreadsheet program to start with, but it's best to create "sanity check" scripts that will go through all the data points and test your assumptions. Ideally, the outputs of these scripts should be binary ("pass" or "don't pass"), but it's okay if they show you a bunch of values to check manually. You can also have these scripts run automatically when updating your model with new data, in the same way that you would run Continuous Integration tests when updating a web application.

Scripts are also used to clean up data. When you don't have too much of it, you can clean it up manually in Excel, but if you have 50MB or more of data and you work on a small laptop, this can turn into a nightmare. In these cases, I highly recommend using a data-wrangling tool such as Pandas[11] in Python (check out the 10-minute tutorial[12] and the 10-minute screencast[13]). Below is a sample script to give you a flavor of Pandas:

```
import pandas as pd

df = pd.read_csv('duplicates.csv', index_col=0,
encoding='utf8') # this creates a Pandas
"DataFrame"

df.drop_duplicates(inplace=True)

# Remove rows where one of the specified
columns is NaN

df = df.dropna(subset=['a_column',
'another_column'], how='any')

# Sort by date (assuming there is a 'date'
column)

df['date'] = pd.to_datetime(df['date'])
df = df.sort(columns='date')

# Randomly sample 10% of the data

import random
num_rows = int(len(df.index)*0.1) # number of
rows we'll keep: 0.1 times total number of rows
(10%)
rows = random.sample(df.index, num_rows) #
sample num_rows elements of the dataframe index
df = df.ix[rows]

# Save result

df.to_csv('data.csv', encoding='utf8')
```

REPRESENTATIVENESS OF DATA

Saying that the training data is *representative* means that it has the same properties as the test data, which is the data on which you will make predictions. For instance, the counts of classes in a classification problem should reflect natural partitions of objects as they occur in the world, and should be the same for training and test data. If you are classifying men vs women, you should observe 50% of each in both training and test data. If you want to detect frauds, the percentage of frauds should be small. If you want to detect important emails, they should occur about 10% of the time.

In maths-speak, we say that the test data should have the same *distribution* as the training data. Without going into probability theory, this means that if you would plot both datasets (as we did in Chapter 2) you would see similar clouds of points. Another way to think about this is that objects pop up in the training dataset in the same way as they do when we request predictions in our application. This is particularly important when estimating the performance of a model before its deployment into production, as we will see in the next section.

I mentioned in the previous section that there could be times when you might want to use data that's been collected by someone else, in order to create a predictive model that you're going to use on new data that *you* will come across. In light of the importance of the representativeness of training data, you should be wondering how the data was collected by that someone. There are cases where this will make you realize that you need to collect your own data. Let's say, for instance, that I have created a dataset of all the emails I have received, classified by importance. You'll agree that it would be a bad idea for you to use this as your training data for building a predictive system of the importance of your own emails, because you and I probably have different criteria for assessing the importance of emails.

You should also be looking out for biases in the data collection process. For house price data, if we are only getting transactions from one website, then there's a bias towards the type of properties sold by the people who use that website (who may be of a particular social group, for instance).

You've now made every effort possible to ensure that your training data is going to be representative of new data, but how do you actually check that it is the case? Let's say that you assign a label ("A") to the objects you have in training, and you assign another label ("B") to the new objects on which you query your predictive model. The trick is to consider a new classification problem where you try to predict that new label ("A" or "B"). If you can create a predictor that has significantly better accuracy (i.e., percentage of correctness) than a random predictor, it means that you have found a way to discriminate between the training data and the new data, hence that the training data is not representative.

An ML system can only be **as good as the data** it was given to train on.

ANONYMIZATION AND PRIVACY

When dealing with data on your users or your clients, you should pay particular attention to data privacy. You shouldn't just send that data as it is to Prediction APIs (or any other third party), and should anonymize any personal references beforehand. The best way to do this is with hash functions that transform a string into a seemingly random string. For example, you should hash email addresses, so that "louis@dorard.me" is represented by

"ecf84b662c041290c62b91c34db9e8991312368c90b9717ab3dac90f34582923"

when using SHA-2 (which is probably the most popular hash function today). However, if someone can get access to your data and is looking for the information of a particular person, he can just hash the email of that person and see if he finds a match in the data. To prevent this sort of targeted attack, you need to add a "salt" to the string before hashing it (so, you would hash "louis@dorard.mesalt" if your salt is "salt"). Of course, if the attacker also knows the salt, you're in trouble, but it requires the attacker to access both your data and the code where you do the hashing and specify the salt. This is a harder task.

Speaking of privacy, are your users aware of what you are doing with their personal data? Typically, this information should be in an easy-to-read Privacy Policy document that is easily accessible from your website/app. Also, if you record data on what your users do with your application, are they aware of it? Even though it would result in less data, you should give them the option to not be tracked. Users are becoming increasingly concerned with their data and privacy. The last thing you want is people leaving your app and giving you bad publicity because of a poor privacy policy. And if you think they won't notice and they won't care, check out Wil Wheaton's experience[14].

MEASURING PERFORMANCE

"That which is measured improves. That which is measured and reported improves exponentially." — Karl Pearson

The performance of an ML system can be measured along several axes: the resources (time) taken for training a model and for making predictions with it, the quality of the model through the accuracy of the predictions, and any domain-level consequences to using the ML system. In this section, I will focus on the last two axes of performance measurement.

PREDICTIONS' ACCURACY

The first thing you can measure when introducing predictions in your app or business is the accuracy of predictions. For this, you need to be able to know, after making a prediction, what the true output value should have been. This would be done in the same way as when collecting training data and getting output values for the example inputs.

Measuring the accuracy of a regression model is pretty straightforward: you look at differences between the truth and what the model predicted, which are called the *errors*. If it is as bad to over-predict as to under-predict, these differences can be combined by taking the mean of the absolute errors (*MAE*).

For binary classification, you would count the number of correct predictions and divide it by the total number of predictions, which is simply called *accuracy* and can be given as a number between 0 and 1 or as a percentage. The accuracy is very simple to compute and to understand, but it is not a very useful performance measure when the classes are imbalanced, such as when you want to detect rare diseases or frauds. Imagine that the most common class, called the *negative* class, occurs 99% of the time. You will have 99% accuracy by always predicting the negative class! You'll agree that, even though it has almost perfect accuracy, this is a terrible predictor. Here are some other performance measures you could be looking to maximize (all between 0 and 1 — the higher the better).

- *Recall* is the proportion of positive objects that were indeed detected as positive by the model.

  - For two models that have same accuracy, Recall will be higher for the model that detects the more positive objects.

  - Recall is equal to 0 when no positives were detected as positives.

  - Recall is equal to 1 when all positives were detected as positives.

- *Precision* is the proportion of actual positive objects out of those who were detected as positive by the model.

- For two models that correctly detect the same number of positives (i.e., that have same Recall), Precision will be higher for the model that detects the less negatives as positives. As a consequence, measuring Recall is a way to check "coverage" of the positive class, while measuring Precision is a way to check that models are not being too loose about what a positive is (ideally they should detect the actual positives and no more).

- Precision is equal to 0 when all positives were detected as negatives.

- Precision is equal to 1 when all negatives were detected as negatives.

- The *F-measure* combines Precision and Recall in one single measure (it is the balanced harmonic mean of the two).

- For two models that have the same Recall, the F-measure will be higher for the model that has higher Precision. Respectively, for two models that have the same Precision, the F-measure will be higher for the model that has higher Recall.

- The F-measure is equal to 0 when no positives were detected as positives by the model, or when all positives were detected as negatives (the two propositions are equivalent).

- The F-measure is equal to 1 when the model has 100% accuracy, which is when all positives were detected as positives and when all negatives were detected as negatives.

I am not including the formulae that define these measures in order to not get you bogged down in mathematical details, but if you're interested, you can find them on Wikipedia[15]. Not all Prediction APIs can evaluate your models on these measures, yet you'll find open source implementations on the Internet for your language of choice (if using Python, check out Scikit-learn[16]). I strongly advise to use open source implementations rather than your own. This is not for performance issues, but because I know from experience that, in this sort of code, it is very easy to introduce mistakes that end up being only visible to a fresh eye.

Alternatively, you can just look at the counts of True/False Positives/Negatives to assess the performance of a classifier.

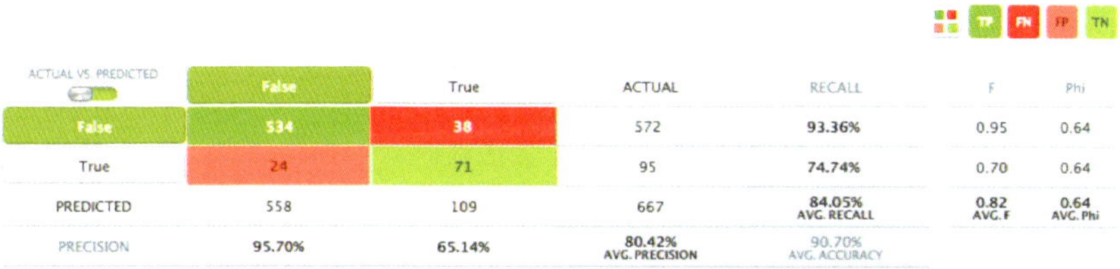

TP FN FP TN

ACTUAL VS. PREDICTED	False	True	ACTUAL	RECALL	F	Phi
False	534	38	572	93.36%	0.95	0.64
True	24	71	95	74.74%	0.70	0.64
PREDICTED	558	109	667	84.05% AVG. RECALL	0.82 AVG. F	0.64 AVG. Phi
PRECISION	95.70%	65.14%	80.42% AVG. PRECISION	90.70% AVG. ACCURACY		

Example of a "confusion matrix" on BigML, giving the counts of true/false positives/negatives

ESTIMATING PERFORMANCE IN ADVANCE

Before deploying a model into production, you need to have an estimate of how well it will perform. The first thing that comes to mind is to evaluate the performance of the model on its training data, by asking for predictions on the training inputs, comparing the predictions to the actual outputs, and computing performance measures. But you'll agree that we could easily create an algorithm that has perfect accuracy on the training inputs by just storing the training examples and, when asked for a prediction for a known input, by returning the output that was associated to it. ML algorithms won't do that, but still, evaluating on training data is a bit like cheating.

A better way to estimate a model's performance is to split your training data into two subsets: one that you will actually use for training, and the other one that you will use to evaluate the trained model. Let's call them the training set and the test set. The idea is that you know the outputs that should be associated with the test inputs, so you can compare them to the outputs predicted by the model you've trained. Before splitting your data, it's best to shuffle it randomly (while still keeping input-output associations). Sometimes you'll realize that the data you're working on has been sorted according to the value of a certain feature, hence a split could skew the training set towards certain feature values and thus wouldn't be representative.

A priori, you would make a 50/50 split. However, it's common to do 80/20 or 90/10 (for training/test), so the models you evaluate are built from almost as much training data as the one you'll really be using (which will have 100% of the training data). It could be that, by chance, the remaining 10% used for tests contain data points for which predictions will be very easy to make (or, conversely, very difficult to make). As a consequence, it's best to do several evaluations with different splits of the data, where 10% of the data is randomly taken as a test set and the rest is used to build a model. You then average the results of the evaluations.

This final technique is called Cross-Validation and can be used with any custom-made performance measure to estimate the performance of a model before its deployment in production. You will need to run the Cross-Validation procedure yourself, but you can find open source implementations. I would recommend Scikit-learn's ShuffleSplit[17], in Python.

BASELINES TO PUT THINGS IN PERSPECTIVE

> *"You don't need to predict accurately to get great value. Organizations, by making predictions per person that are a good bit better than guessing, actually play the numbers game better."* — Eric Siegel

Let's say, for example, that you're interested in the accuracy of a classification model, and that your model has an estimated 80% accuracy. Is it any good? The first thing to do to find out is to compare to a baseline. For classification, the baseline could be:

- A random classifier.
- The mode classifier, which always returns the class that was found to be the most common from training (called the *mode*); in binary classification, this is the negative class.

Our performance measures for the mode and random classifiers have some interesting properties:

- Accuracy

- Mode: equal to the proportion of the most common class

- Random: about 50%

- Recall

- Mode: 0%

- Random: about 50%

- Precision

- Mode: undefined (because it predicts no positives)

- Random: about 50%

- F-measure

- Random: about 50%.

There's another performance measure called *Phi,* which is similar to the F-measure in that it combines Precision and Recall, but in a way such that Phi is about 0 for the random classifier.

For regression, the baseline could be:

- A random regressor that returns a number at random in the range of output values found in training.

- The mean-value regressor that always returns the mean output value as computed from the training data.

The *R-squared* value of a regression model indicates how much better it is than the mean-value regressor. It is simply equal to one minus the ratio between the *MSE* (Mean Squared Error) of the model and the MSE of the mean-value regressor. The better the model is (in comparison to the mean-value regressor), the closer the value of R-squared is to one. Note that R-squared can take negative values if we're doing worse than the mean regressor.

Now, you've compared your model to a baseline and you've made sure it was better, but still, the question remains: is 80% accuracy any good? It all depends on your domain of application, and the question may be hard to answer because you may not be looking at a performance measure that has a meaning in the context of your application. This is why you should also measure the improvement provided by your predictive model over a baseline predictor using the real-world performance measures that you defined for your particular application, as we discussed at the end of the first section of this chapter ("Making predictions... so what?").

Measuring the accuracy of predictions is good.

Measuring the impact on your objectives is better.

CHAPTER 6: PREDICTION APIS

6

WHICH APIS?

In this book I have chosen to only focus on Prediction APIs...

- with a fixed pricing structure that allows for bootstrapping (free trial or price plan affordable by individuals);
- to which you can sign up instantly;
- in which there are no algorithmic parameters to tune (which excludes Algorithms APIs).

I'll start with a section dedicated to Specialized Prediction APIs. Then, there will be one section dedicated to each of the main two Generic Prediction APIs: BigML and Google Prediction API. Finally, I will mention a few other Generic Prediction APIs.

Please bear in mind throughout this section that the information it contains was only accurate at the time of publication (see first page) and that things may evolve. You should check the websites of the various services presented here for any changes to their capabilities and their pricing.

SPECIALIZED PREDICTION APIS

Specialized Prediction APIs realize predictive tasks for a specific problem. The provider of the API has already trained a model for the task at hand. There is only one core method to the API, which allows you to specify an input and get a predicted output back. As I said in Chapter 3, Specialized Prediction APIs should be your first port of call after formulating the question you want to predict answers for. Some of these APIs are quite advanced because they are based on ML techniques that are not covered in this book, and they could not be easily reimplemented with existing Generic Prediction APIs. For more examples, check out a list of 40+ Specialized Prediction APIs on Mashape[18].

TEXT / NATURAL LANGUAGE PROCESSING

Let's review some APIs for classification tasks that have text as input. Some of the APIs would use text representations that capture the structure and meaning of the ideas expressed in "natural language" in the text. You'll want to use an API that has been made specifically for the language of the text you will be sending (the underlying training data, hence the model, would be different for different languages).

Let's first have a look at APIs where the input text represents a document, post, webpage, or message. Typical use cases are on user-generated content or content crawled from the web.

- Alchemy[19] has an API for sentiment analysis (as introduced in the Examples chapter), for language detection and text categorization, which can deal with HTML or web-based content and classifies content into categories such as "news", "business", "sports", etc.

- Datumbox[20] has competing APIs that can also specifically target tweets for sentiment analysis, and categorize webpages into "commercial," "educational," "adult-only." In addition, they have a spam detector (for email or comments) and a readability assessor which classifies documents into "basic," "intermediate" and "advanced."

- Semantria[21] is another API provider that provides an Excel plugin so that predictions can be made from within spreadsheets. They report that their clients use the API on customers' ratings of products and comments to better understand what they think about the business and to find out where problems exist.

In other APIs, the input text represents a query, as with Maluuba[22]. Its API can be seen as a "Siri for developers." You send a query that a user of your app formulated in natural language (e.g. "Get me a flight from New York City to San Francisco for next Thursday", "What awesome action movies can I get 3 tickets to for tonight?", "Remind me to do the laundry in 2 hours", etc.) and it returns a type of query (e.g. "reminder_set") along with parameters for that type of query (e.g. "message": "do the laundry", "time": 2013-12-16 11.12am). You then use this as a structured representation of the query that your app can act upon.

There are actually two APIs at play in Maluuba that solve two different problems: the Interpret API classifies phrases into predefined types of queries (e.g., "reminder_set", "travel_flight", "entertainment_movie", "calendar_create_event", etc.) and the Normalize API finds normalized values for dates and times expressed in natural language (e.g., "tomorrow").

COMPUTER VISION

Another type of powerful APIs is those that deal with images as inputs. Some specialized processing and feature engineering are usually needed to deal with images in an ML system, and all of that is done for you by the API. The examples that follow are, again, classification tasks.

Alchemy, which we just covered in the previous subsection, also has an API for detecting objects within images, but it is not yet open to the public. The API takes images in input and performs multi-class/multi-label classification (i.e., it outputs more than one class). The set of possible classes/labels was defined from the data used for training. They can be, for instance, "dog", "car", "pen",... well, the list is quite long!

Nsure.io's Porn Filter API[23] scans images to determine if they contain pornography by "looking" at skin tones, shapes and other cues in the images.

Ocrapiservice.com performs Optical Character Recognition (OCR). It takes an image of text as input and returns text in the form of a string of characters. In its most simple form, the OCR task is a classification task with an image of only one character in input, and with a character code (in ASCII for instance) in output. Representing the input as just a pixel matrix can work quite well[24] for hand-written characters (also see the Digit Recognizer model in the BigML gallery[25]).

BIGML

All Generic Prediction APIs, including BigML's API[26] and Google Prediction API[27], have two core methods in common: one to feed data (input and output examples), which triggers the learning of a model, and another one to ask for a prediction of an output given an input and a model. But if you remember the code given at the end of Chapter 3, in Python (BigML also provide API wrappers for other popular programming languages), there were actually 4 methods:

- `source = `**`create_source`**`(training_data_csv_file)` and `dataset = `**`create_dataset`**`(source)` to set up the data.

- `model = `**`create_model`**`(dataset)` to learn a model from the data.

- `predicted_output = `**`create_prediction`**`(model, new_input)` to make a prediction against the model.

This is because BigML makes a distinction between a data source (where data comes from) and a dataset (what you actually use to make the machine learn).

You can do the same things on the BigML.com web dashboard. It allows you to upload a spreadsheet or CSV file to define a data source. Alternatively, you can specify a file URL, an Azure[28] source, a Google Docs[29] URL, an Amazon S3[30] object, or a Freebase[31] locator — BigML will read and parse data from there, transforming it to the moral equivalent of a CSV file. BigML can deal with numbers, labels, and text as feature types. It has a built-in text analysis tool that automatically detects the language of a piece of text and then does stemming, stop-word removal, and tokenization (a technique for splitting text into tokens in a smart way, so that "New York" is treated as a single word/token — not as "New" then "York").

Setting up your data and asking for predictions can be done from the web interface without coding, which is great for getting to know the system and also because of the data visualization feature.

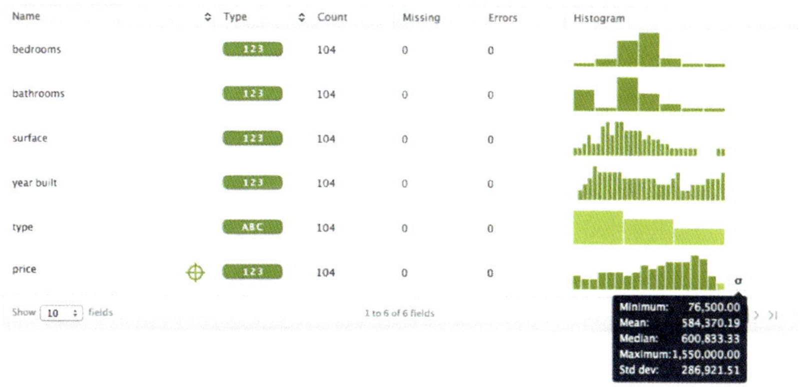

Dataset visualization with histograms

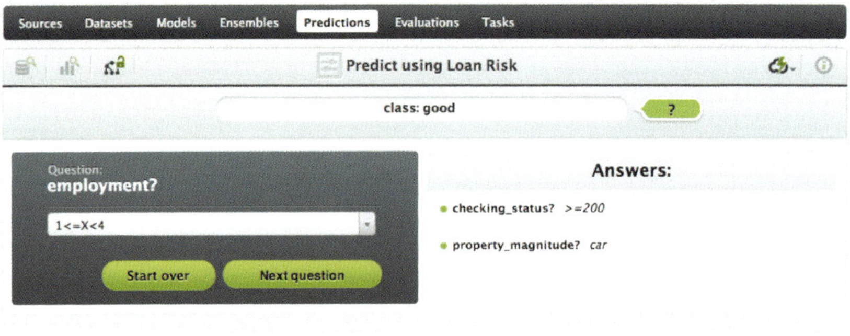

Question-by-question web interface for Loan Risk prediction

Advantages

Before going into the advantages of BigML, I should first explain that it uses "Decision Tree" algorithms to make predictions. A Decision Tree represents a prediction algorithm where each node is associated with a question on the value of an input component, with a certain number of possible answers represented by branches, and where leaves are associated with output values. The first question is located at the root node. Choosing an answer takes you to a branch of the tree and to a next node. The process is repeated until a leaf is reached, where you get the associated output value as a prediction. The structure of the tree, the questions associated to nodes, and the output values associated to leaves are learnt from training data by a Decision Tree learning algorithm.

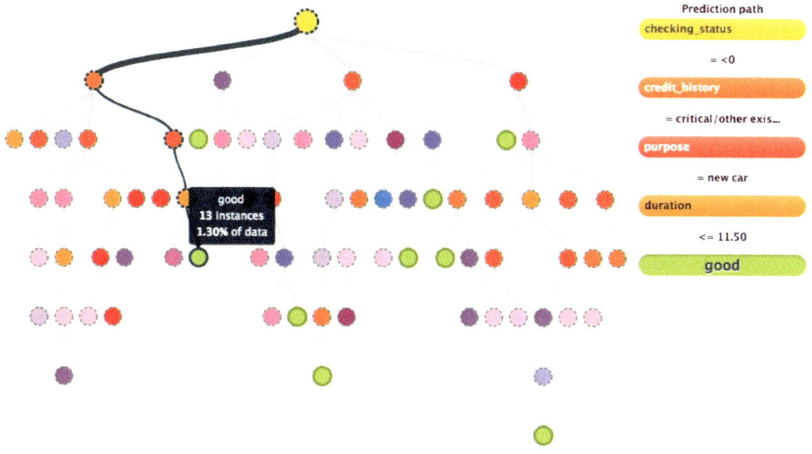

Decision tree that predicts the quality of a loan application

The trees shown on the BigML.com web interface are interactive (see for instance the loan risk decision tree on the BigML gallery[32]). They show and hide branches intelligently so that the tree fits in the browser window while showing what you are interested in. Predictions can be given at both leaves and nodes (but with less confidence). This visualization makes it possible to understand how predictions are made and thus gain insights into the problem. To illustrate this, I recommend you browse BigML's gallery[33] of examples which is organized into a variety of categories. It doesn't require a BigML user account to be accessed.

One advantage of Decision Trees is that they can deal with missing values in the data and make predictions very quick to compute, which is particularly interesting when you have a lot to make. When the training has been done, you know exactly what the Decision Tree is, and thus what the prediction algorithm is. So, you can store it locally (on your server or on a client device) and make predictions offline, without having to connect to an API. The only functionality that requires connection to BigML is the learning algorithm that creates the Decision Tree based on the training data.

Limitations

BigML may not have the same flexibility as Google Prediction API because it is restricted to a certain family of algorithms. It is worth noting that there are problems for which Decision Trees won't perform so well, but characterizing them is beyond the scope of this book. You just need to try and compare to what you'd get using other Prediction APIs.

Another limitation is that you can't stream data to a BigML model, meaning you can't update the model/Decision Tree continuously as you get new data. Instead, you need to create new models from data including any new data.

BigML accepts files of up to 64GB (or 5TB if using Amazon S3) as data sources. As you see, this is way more than what you could use in a spreadsheet program (a few tens of MB on my laptop) or than what would fit on your computer.

Pricing

For those who are just getting started and exploring the service, BigML offers a free plan with an unlimited number of "tasks" under 16MB in "development mode". A task can be a dataset creation, model creation, model evaluation, or a batch of predictions. Up to 4 tasks can be run in parallel with the free plan. The development mode, as opposed to the "production mode," has mild limitations on your usage of the service: the maximum number of terms in text analysis is limited to 32 and the maximum number of nodes in a tree cannot be higher than 512.

The free plan is ideal for trying out ideas and, for some applications, could even work in a production context. If you have bigger needs, you can either pay as you go or get a subscription plan. The starter plan is $30/month and includes an unlimited number of tasks of up to 64MB, with up to 2 in parallel. The biggest price plan raises the limit on the size of the dataset to 4GB (with up to 8 tasks in parallel) but costs ten times as much. It is unlikely that you will need more than what the starter plan offers in the first few months of using the service.

With pay-as-you-go, you buy credits that allow you to perform tasks. The number of credits that each task consumes depends on the size of the data for this task or on the number of predictions made. Which offer you choose in the end really depends on your application, but if you've followed the methodology outlined in this book, you'll be able to estimate the tasks and predictions you will be doing in a given month. Thus, you'll know how much you would end up paying per month with pay-as-you-go or with a subscription.

There is no Service Level Agreement by default with these offers. If you need one (i.e., guarantees on availability/uptime, disaster recovery, or performance), this needs to be discussed individually and the SLA would be tailored to your needs.

Alternatively, if you need to take your application of ML to the next level, BigML has a Virtual Private Cloud solution. This is worthwhile if you have particular requirements in terms of data security, privacy, or uptime. Because you would have your own cloud servers residing within Amazon Web Services, BigML would just be an extension of your own datacenter, and you would benefit from the level of uptime and security of the Amazon platform.

GOOGLE PREDICTION API

As with BigML, you can find Google Prediction API[34] (a.k.a. Gpred) bindings for Python and other languages. I find Gpred particularly interesting to use through Google Apps Scripts[35] (a.k.a. Gscripts) in conjunction with Google Spreadsheets[36] (a.k.a. Gsheets). I have already talked about the importance of seeing your data in a spreadsheet program, and I quite like being able to get predictions for missing values without having to leave the spreadsheet program. I can have performance measures show up in the spreadsheet and update in real-time as predictions are being made. Gpred doesn't automatically give histograms of your data like BigML does, but you could always make them from within Gsheets.

Here is an example in Gscript (which is a language derived from and virtually identical to Javascript) of how you would get data from a Gsheets document:

```
var instances =
SpreadsheetApp.getActiveSheet().getActiveSelect
ion().getValues();
```

This could then be used to create training instances in the format expected by Gpred, as an array of examples made of objects that each contain an `output` field and a `CSVInstance` field representing the array of associated input components:

```
var trainingInstances = [];
for (var i = 0; i < instances.length; ++i) {
  var output = instances[i][0];
  var CSVInstance = instances[i].slice(1);
  trainingInstances.push({'output': output,
'CSVInstance': CSVInstance});
}
```

When this has been defined, a model with name '`modelName`' is created by calling the API as follows:

```
Prediction.Trainedmodels.insert({'id':
'modelName', 'trainingInstances':
trainingInstances}, projectNumber);
```

The model is built asynchronously and can then be invoked by its name when making a prediction against it for a `newInput` array:

```
var prediction =
Prediction.Trainedmodels.predict({'input':
{'CSVInstance': newInput}}, projectNumber,
'modelName');
```

When you're not using Gsheets but are using training data stored in a CSV file, you would have to upload it to a Google Cloud Storage bucket. This can be done from the command line or from the web with the Google Cloud Console[37] interface (a.k.a. Google Developers Console). The uploaded file and its bucket can then be referenced in a parameter of the **insert** method of the API.

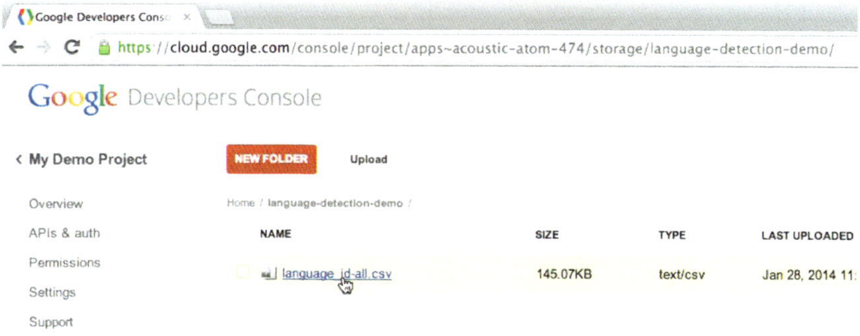

The Gpred functionalities can also be accessed from a web interface, Google APIs Explorer[38]. As you can tell, this is a web-based client for Google's APIs, including Gpred. It's not as user-friendly as BigML's web interface, but still you can upload your datasets, create models and ask for predictions without having to code anything.

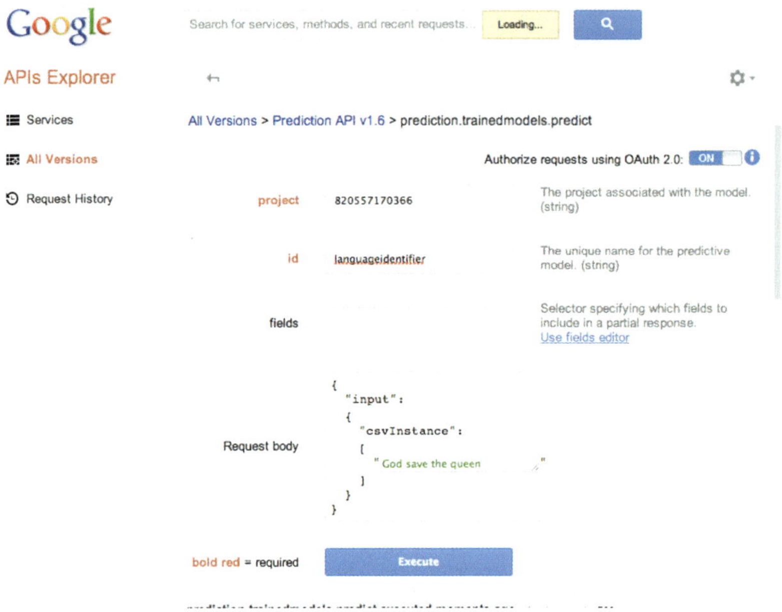

This is more or less all there is to using Gpred. You'll notice that it has less options to play with than BigML. For instance, you won't know how predictions are made and you won't be given a model you can interpret — the service is completely black box.

Advantages

To me, the strongest selling point is having access to Google's set of ML algorithms to power your predictions. Gpred does not restrict you to any particular family of algorithms like BigML does with Decision Trees, and the type of algorithm to use is automatically determined from the characteristics of your training data. Besides, it's very convenient to be able to just throw in more training data, whenever available, via the **update** method of the API, and have subsequent predictions made from a model that uses the freshest data. Note that the update method takes in one data point at a time.

The other advantage of Gpred is how it integrates with Gscripts and other Google services. If your app is hosted on the Google App Engine, chances are that you're already using Google Storage. Also, data transfers between Google servers should be quick.

Limitations

The main drawback of Gpred is that it cannot deal with missing feature values. It won't fail, but it'll evaluate the empty value as the empty string ("") for text and labels, and as 0 for numerical features. The types of features are, as with BigML, numbers, labels and text, but there is no built-in text analysis tool. Instead, you need to use libraries such as NLTK[39] in Python and preprocess your textual data before uploading it to Gpred.

The maximum size of the CSV files you can send to the API is 2.5GB.

Pricing

First of all, note that there are two costs to consider when using Gpred: the Google Prediction API itself and, if uploading CSV files of training data, Google Cloud Storage. The latter would cost you less than 50 cents per month for a 2.5GB file — check out the pricing online for more details[40].

Regarding the Prediction API itself, there is a free quota during the first 6 months of using the service in a given project. It includes:

- 100 predictions per day.
- 5MB of training per day.
- 100 streaming updates per day.
- 20,000 predictions at most (throughout the whole 6 months).

This is fine for testing the service and trying out some ideas on your initial dataset (or on a subset). Then, paid usage can lift these limits and add a 99.9% availability Service Level Agreement. There is a base fee of $10 per month covering up to 10,000 predictions and 10,000 streaming updates (see the pricing rules[41] if you anticipate you'll be doing more). You should then add $0.002 per MB of data used for training via the **insert** method (which makes about $5 for a 2.5GB dataset).

OTHER GENERIC PREDICTION APIS

Classification and regression

uClassify[42] deals with text classification problems, meaning that the input is always a piece of text and the output is a class. This is more restrictive than Google Prediction API and BigML, but the fact that you can define which are the possible classes and provide your own training data did make it the first Generic Prediction API when it launched in 2011.

Predictobot[43] (in private beta) presents prediction problems in the same way as we did in the spreadsheet view of Chapter 3. It fills in missing values in a spreadsheet, one column at a time. By looking at existing values (labels or numbers) in the column in which to predict missing values, it automatically knows whether it's a regression or classification problem.

Predictobot is actually not a Prediction API per se. You'll upload a spreadsheet (Excel, CSV, etc.) and get another spreadsheet back, which is a copy of the original one with an integrated formula for predicting new values. You can just apply the formula (as with any other formula in a spreadsheet program) to empty cells in the target column of your spreadsheet. There's no coding involved here, and since most people can use a spreadsheet program, this service should be very accessible.

Working with spreadsheets is fine as long as it's reasonable for your application to stay within a spreadsheet program. This can work as a first test of the predictions you could do with the data you have. Otherwise, it would restrict you to certain types of business applications (for apps, you would need programmatic predictions). Predictobot also tells you which features are given the most importance in the model, which is useful to get some insights.

Natural Language Processing

Wit.ai more or less does the same thing as Maluuba (the "Siri for developers"), but it allows you to define your own types of queries (called "intents") that have their own sets of parameters (called "entities"). The vision behind this service is to replace traditional user interfaces with text- or speech-based ones, where the user just types text into a box or speaks a command which expresses what it is he wants to do in natural language. This is of particular interest in hands-free mobile usage (when cooking or driving, for instance) or in wearable devices that have tiny screens and minimal controls. For this to work, Wit.ai needs to train a model from examples, provided by you, of queries as they could be expressed by your users (the inputs of the model, also called "expressions") and the associated intents (the outputs).

CHAPTER 7: CASE STUDY - PRIORITY INBOX

7

SPECIFYING THE PROBLEM

We first covered Priority Inbox in the Apps examples of Chapter 4. Here, I'll show you how I would reimplement the system that predicts the importance of non-spam emails. For this, I will apply the recommendations given in Chapter 5. Let's assume that all emails in a thread started by the user are always important. For every incoming email starting a new thread, we will try to predict the answer to the following question: "Is this email important?" For emails belonging to a previous thread, let's assume that the importance is the same as the importance of the first email in the thread.

Google Apps Script[44] is a platform for running and editing scripts on Google servers that can access your personal data and documents on Google services. I will use it to access my emails on Gmail and collect data to train a model on. Inputs are the first emails of threads found in the inbox. I will use the `isImportant()` method of the GmailThread class[45] to get corresponding output values. Note that the value this method returns is either Gmail's opinion on the importance of the thread, or it comes from explicit feedback given by me through the Gmail web client.

To evaluate the performance of the system, I will sort the emails in the order I received them, and I'll make a 50/50 split: I will use the first half for training a model with BigML, and the second half for testing the model.

Before any work goes into the creation of a model, we need a deeper understanding of the problem, a vision of what the predictions will be used for, an idea of what value they will bring, and a concrete objective to achieve with them.

Gmail meter[46] is a good tool for getting insights into how one uses emails (see also "Know your Gmail stats using Gmail meter[47]"). Configure it with your email aliases, fire it up, authorize it, and wait for the report which will be sent by email. I noticed that the top 5 email recipients change from one month to the other, but they are always in my address book. The top 5 senders were mostly addresses that send unimportant notifications, and addresses that send rather important notifications (that I tend to read). I'm not sure yet how it can help, but I'll keep that in mind when engineering features. I also learned that I receive an average of about 30 emails starting new conversations per day (for some people it could be ten times more) and that about 10% of all conversations are important. This makes 3 new important conversations per day on average.

I want to use predictions to spend less time on email. Ideally, I want to check emails once a day and only focus on the important ones. They should be grouped together in a section at the top of my inbox (see illustration given in Chapter 4). Then, I would also do longer email sessions twice per week. In these sessions, I'd go through the whole list of emails in my inbox and treat all those I received since the last session. I'll plan, as an initial objective per day and on average, to detect no more than 1 email that's actually not important (False Positive). Important emails that were not detected (False Negatives) are problematic since it takes me more time to take action on them, since I'll need to wait until the next bi-weekly email session to be aware of them. Let's fix an objective of 1 FN every 3 days. These objectives are somewhat arbitrary, but the reflection that leads to them is important. What's for sure in the end is that FNs are even less desirable than FPs.

FEATURE ENGINEERING

BACKGROUND REVIEW AND INSIGHTS

The importance detection problem reminds us of another email classification problem: spam detection. In the popular Spambase Data Set[48] you would find that most features are based on counts of capital letters. In an example of the Google Prediction API Developer's Guide[49], we are told, "In the spam detection model, we could include the number of external links in the email, the number of images, and the number of large attachments." The set of features used to characterize emails and which would have an impact on their importance can be expected to be different, though. Let's review some prior work to give us some ideas. In "The Learning Behind Gmail Priority Inbox" (2010)[50], Aberdeen et al. report using "many hundred features" which they categorize in:

- **Social features**, based on the degree of interaction between sender and recipient.
- **Content features**.
- **Label features**.

Here is some more information found on the Internet:

- Priority Inbox overview (Gmail Help)[51]:
  "Gmail uses a variety of signals to prioritize your incoming messages, including who you've emailed and chatted with most and which keywords appear frequently in the messages you opened recently."

- Gmail's Priority Inbox Is Awesome. Just Give It a Chance (Whiston Gordon, on lifehacker.com, 27/3/2013)[52]:
  "Priority Inbox learns, first and foremost, just by watching you. If you frequently read messages from a specific sender, it learns to mark those as important, and if you frequently reply to emails from them, it knows even better. If you delete stuff without reading it, it'll learn that those aren't important."

  "I have a filter that applies a label called "Internal" to emails coming from any of my coworkers. Gmail now recognizes that many Internal emails are important, though it doesn't always mark them as important. Depending on the sender and the context of the message, it will use the Internal label as one more deciding factor, which is really handy."

The Gmail interface also allows you to see the reasons emails were marked as important when hovering on the yellow importance icon. These read: "Important mainly because..."

- It was sent directly to you.

- Of the people in the conversation.

- It matched one of your importance filters.

Another thing I did was search for important threads that I didn't start, by typing the following query in the Gmail search bar: `"-from:louis@dorard.me label:important"`. It looks like the results were emails sent by friends or people I communicate with frequently, or whose emails I always read. Or, they were sent by people I've sent emails to in the past, or were sent by people who are usually in important conversations.

LIST OF FEATURES

Based on all the previous information, I chose to extract the following features from my emails:

- **Subject (text)**.
- **Sender's email address (categorical)**.
- **Alias (categorical)**.
- **Me in the email recipients (categorical)**: "only" if email sent only to me, "to" if sent to me with other people in cc:, "cc" if I'm in cc:, "bcc" if I'm not in cc.
- **Number of attachments (positive number)**.
- **Body/content (text)**.
- **Length of body (positive number)**, which value is the number of characters of the value of the previous feature.
- **Concatenated list of email labels (text)**, where labels are separated by a single space.
- **Number of conversations** with the sender/contact: the number is arbitrarily capped at 20 (for faster processing and extraction of the feature values).

- **Proportion of threads started by contact (number between 0 and 1)**.

- **Proportion of threads read**, i.e., with all emails read in the thread.

- **Proportion of threads starred**.

- **Proportion of emails contact replied to**.

- **Average time taken by contact to reply (number in milliseconds)** to first email in thread started by me. If this contact has never replied to my emails, the value will be infinity, which I'll replace with a big number (I have chosen 2^64-1, i.e., the biggest integer value in 64 bits).

- **Average length of contact's reply (number of characters)**.

- **Proportion of emails I replied to**.

- **Average time taken by me to reply** to first email in thread started by the contact.

- **Average length of my reply**. If my reply is long, then I care about the contact's emails.

- **Contact in address book (categorical)**: "True" or "false."

See how I chose to use "normalized" features when possible, such as proportions, in order to facilitate the learning. You could come up with more features and I haven't encoded all the ideas mentioned in the previous subsection. I'll leave it as an exercise: see if you can get a better model than me!

PREPARING THE DATA

COLLECTING EMAIL DATA WITH GOOGLE APPS SCRIPTS

I used the search method of the GmailApp class[53] to collect all messages that were not spam within a fixed time period. I paginated the results, looped over them, and extracted the required information. Calls to the Gmail Service[54] and to the Contacts Service[55] are quite long. It takes about 1 minute to extract the info related to a thread and its sender! Besides, the maximum running time of a function in Google Apps Scripts is 6 minutes and there is a restriction on the number of consecutive calls to the different services (you can check out your quota online[56]).

So, I created a function that only retrieves 5 emails, along with a trigger to run this function every 5 minutes — it did the trick! The information I extracted was stored in a ScriptDb[57] data structure, which also has restrictions in terms of how much data you can store. I then dumped features into a Gsheet along with three more columns: thread id, email date and time, and email importance. Because I am only considering the first email in each thread, the values in the thread id column are unique.

In total I collected 4,710 emails over a period of 5 months. I downloaded the Gsheet as a CSV file of 9.27 MB. Let's take half of this for training and let's make sure, based on the guidelines given in Chapter 5, that we have enough training data. We have 19 features and 2,354 data points, which is 124 times as many (a factor 10 is a minimum, and 100 or more is recommended). There are 2 classes, important and regular, with 207 important emails (> 10). We should now check that we have at least 10 instances per category (ideally a few hundred) for each categorical feature:

- Me: "only" (1921 instances), "to" (41) , "cc" (40), "bcc" (352).

- Sender: there are 458 senders, so on average there are 5 emails per sender (but this is just an average). Ideally, we should collect more data.

- Contact in address book: "true" (368 instances), "false" (1986).

Id	date	subject	sender	...	In address book	important
13c009bd7	2013-01-03 07:31:18	Birthday!	johndoe@gmail.com		1	1
13c009d94	2013-01-03 07:43:12	Daily digest	daily@hojoki.com		0	0
13c009e1f	2013-01-03 09:09:48	[JDW6] Prenez date...	camille@neocamino.com		1	0
13c009e35	2013-01-03 08:33:53	New deals in Paris	info@groupon.com		0	0
13c00a6fc	2013-01-03 09:12:29	Recsys results	quentin@codole.com		1	1

```
id, date, subject, sender, in address book, important
13c009bd72f9a147, 2013-01-03 07:31:18, "Birthday!", "johndoe@gmail.com",
TRUE, TRUE
13c009d947d77197, 2013-01-03 07:43:12, "Daily digest",
"daily@hojoki.com", FALSE, FALSE
13c009e1fc1db460, 2013-01-03 09:09:48, "[JDW6] Prenez date...",
"camille@neocamino.com", TRUE, FALSE
13c009e3587b25c2, 2013-01-03 08:33:53, "New deals in Paris",
"info@groupon.com", FALSE, FALSE
13c00a6fc597d204, 2013-01-03 09:12:29, "Recsys results",
"quentin@codole.com", TRUE, TRUE
```

Excerpt of what an email data spreadsheet and its corresponding CSV file would look like

SANITY CHECKS

Let's fire up Excel to browse the data and perform some basic sanity checks by sorting and filtering the column values. In order to keep Excel responsive, I used it on 20% of the data at first. I formatted the spreadsheet so the data would be nicer to look at (converting proportions to percentages and rounding off numbers). While exploring the data, I found it useful to create a header row and freeze it, and move the 'importance' (output), 'subject,' and 'sender' columns to the left and freeze them as well. This let me keep them in sight when scrolling, see relationships between features and output values, and identify emails in order to interpret what I saw. I also temporarily deleted the 'content' column which contains a lot of data so I could manipulate the spreadsheet faster.

As I browsed the data for some general inspection and sorted the values in columns one at a time, I noticed two things:

- For some reason, a couple of output values were 'undefined'; I chose not to investigate this and to just drop the corresponding data points.

- A small number of values were missing for 'length' and 'content'. Apparently, the method from the GmailThread class that gets an email's content can fail with some message types, so there's not much that can be done here. Fortunately, BigML's Decision Trees can deal with missing feature values.

Let's list some basic assumptions that should be verified by the data:

- No NaNs (Not A Number) in the numeric features.
- No empty values except alias which is empty/missing when I am in bcc:. In this case, the data collection script can't find any of my aliases in the recipients.
- 'conversations number' and 'length' should both be strictly positive.
- 'recipient' shouldn't be empty.

When exploring the data to check these assumptions, I noticed a few bugs in my feature extraction and dump scripts. I also realized that some 'label' and 'subject' values were empty and this would be considered by BigML as missing values (which they were not), so I decided to replace the empty values with "nolabel" and "nosubject". I made the appropriate changes to the scripts. I also added a line of code to tidy up strings of text before dumping them. I trimmed them, removed non-word characters and digits, changed all spaces to single spaces, and surrounded them with quotation marks.

```
function tidyText(s) {
  return "\"" + s.replace(/[\W\d]/g,
'').replace(/\s+/g, ' ').trim() + "\"";
  // see http://www.w3schools.com/jsref/
jsref_obj_regexp.asp for more info on regular
expessions
}
```

I just had to rerun the scripts to get a "clean" dataset. There are situations, however, where you would get data as a CSV or Excel file without a script that generated it. You would then have to clean it directly (with Pandas or an equivalent data-wrangling tool).

ASSUMPTIONS TESTING

Let's now test our insights into the email importance problem by listing assumptions on how feature values relate to output values:

- **If, on average, it takes me a short amount of time to reply to the contact, then his emails are likely to be important**. This is almost always true, except for some instances where emails were incorrectly marked as unimportant — typically in newsletters or Forum threads sent by people I know personally.

- **If the 'in address book' feature is true, then the email should be important**. This turns out to be occasionally false, as I noticed emails that were incorrectly marked unimportant although they were important (and they were sent by friends). Similarly, I realized that some senders were in my address book when they shouldn't have been, due to the fact that I had previously experimented with a Gmail Labs feature that would automatically add new senders to my contacts/address book. This made it a mess, and I'm wondering now why they made such a feature — and why I used it!

- **If the 'me' feature is 'bcc' and the sender is not in my address book, then the email is not important** (always verified).

- **Emails with several attachments are likely to be important**. This turned out to be true in most cases, except for some emails that are in a "gray area" (I'm not sure myself if they're important or not), and for some Forum/Google Group emails I get that are not important to me and sometimes have attachments.

- **'%read' or '%starred' high should imply that email is important**. This is true when I have had many conversations with the contact (otherwise this percentage wouldn't have much significance). There were exceptions in a few cases where the output was incorrect.

- **If the average time taken by the contact to reply to me is rather long, then I'm likely to be impatient to read from him**. I realized this wasn't true at all when '%from' was close to 1, such as when almost all of the conversations with the contact were started by him. I filtered out the rows with '%from' values greater than 0.9 and things were much closer to what I expected. I also noticed while filtering rows that all emails with '%from' values lower than 0.7 were important (in my case).

- **If my replies are long, then I care about the contacts' emails**. Again, checking this assumption brought up a few emails that I had seen before and were marked as unimportant although they were, but the assumptions were almost always true.

You can think of more assumptions to check if you like, but I'll leave that as an exercise. From these checks, it turns out that the Gmail importance labels and feature values have a certain degree of noise, but we are able to identify problematic data points and could correct their outputs manually if time and resources permit. Let's leave the data as it is, and see if the noise can be dealt with!

IMPORT INTO BIGML

It's now time to upload the CSV file to BigML, which makes up a new data source. When creating a dataset, we make sure that feature types are correctly identified:

- Subject and content should be text. I have chosen to activate BigML's text analysis functionality and to have it detect the text language automatically and remove stop words.

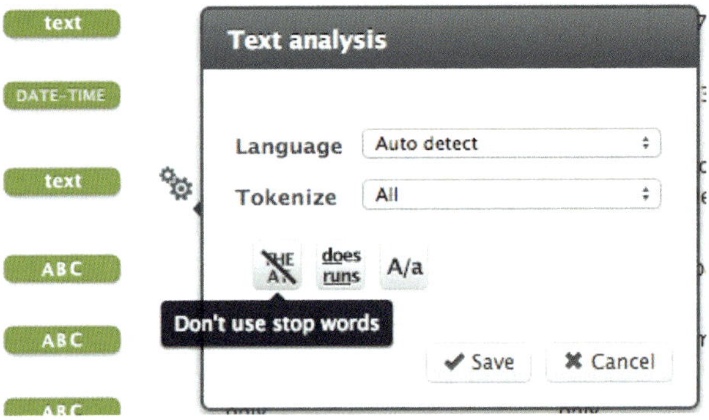

- Labels should also be considered as text (but it is not in a language). The reason is that I am representing the set of labels of an email with a concatenation of these labels, e.g., {"work/project_X", "notification", "followup"} is represented by "work/project_X notification followup".

The dataset is shown in histogram view within the BigML web interface (see next page), which allows us to make a final inspection of the data before creating a model from it. At first sight, the histograms seem to make sense. There are no erroneous values, 9 missing content and length values, and 311 missing alias values. We can also look at statistics of the feature values (like mean, variance, and range) by hovering the mouse on the right-most "sigma" icons (not shown in the screenshot). We see that BigML correctly identified the last column in the source ('importance') as the output (also called *target*), and it automatically created additional date features based on the date column ('month', 'day of month', 'day of week', 'hour', 'minute', 'second'). We could choose to deselect these date features (and not use them for building a model) but we'll keep them to see what happens. The exclamation marks you see next to 'id' and 'date' mean that these features are "not preferred". In other terms, BigML understands that they won't be useful for model training and it won't consider them.

id	!	text	2,354	0	0
date	!	DATE-TIME	2,354	0	0
subject		text	2,354	0	0
sender		text	2,354	0	0
alias		ABC	2,043	311	0
me		ABC	2,354	0	0
attachments		123	2,354	0	0
content		text	2,345	9	0
length		123	2,345	9	0
labels		text	2,354	0	0
conversations		123	2,354	0	0
%from		123	2,354	0	0
%read		123	2,354	0	0
%starred		123	2,354	0	0
% contact replied		123	2,354	0	0
time contact reply		123	2,354	0	0
length contact reply		123	2,354	0	0
% me replied		123	2,354	0	0
time me reply		123	2,354	0	0
length me reply		123	2,354	0	0
in address book		ABC	2,354	0	0
important	⊕	ABC	2,354	0	0
date.year	!	YYYY-MM-DD	2,354	0	0
date.month		YYYY-MM-DD	2,354	0	0
date.day-of-month		YYYY-MM-DD	2,354	0	0
date.day-of-week		M-T-W-T-F-S-S	2,354	0	0
date.hour		HH:MM:SS	2,354	0	0
date.minute		HH:MM:SS	2,354	0	0
date.second		HH:MM:SS	2,354	0	0

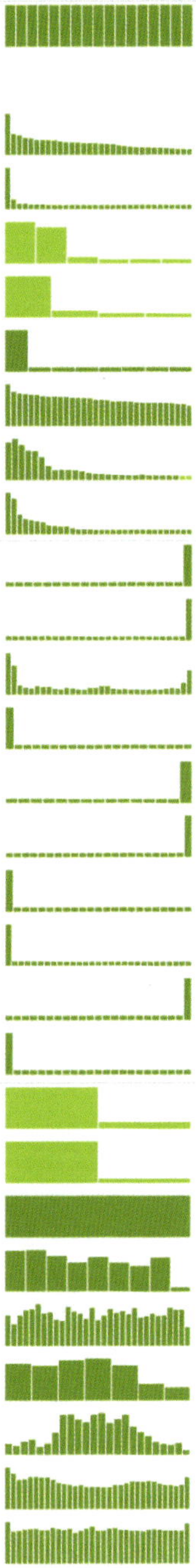

RESULTS

The image below shows the list of tasks I performed in BigML for this case study.

evaluation/52fa86650c0b5e6d4a003099	27min	1.445 secs	●●●●	2K+ inst.	~~23.550~~
model/52fa86410c0b5e6d4d0059c4	27min	2.902 secs	●●●●	4.6 MB	~~18.541~~
dataset/52fa85750c0b5e6d49001e0d	31min	4.771 secs	●●●●	4.6 MB	~~4.634~~
dataset/52fa85750c0b5e6d49001e0a	31min	5.188 secs	●●●●	5.6 MB	~~5.621~~
model/52fa851f0c0b5e6d4a00308c	31min	0.000 secs	●●●●	-9.3 MB	0.000
model/52fa851f0c0b5e6d4a00308c	32min	5.235 secs	●●●●	9.3 MB	~~37.071~~
dataset/52fa83ef0c0b5e6d480020b1	37min	15 secs	●●●●	9.3 MB	~~9.268~~
source/52fa82b70c0b5e6d4d0058ee	43min	14 secs	●●●●	9.3 MB	0.000

There are 5 textual columns in this table:

- Type of task/id of object created.
- How long ago was this task performed. This is useful to identify the task ("Ah, yes, it's the model I created half an hour ago before I did that other thing...")
- How long the task took to complete.
- The size of the task.
- The number of credits consumed by the task. Here, the numbers are struck through since the tasks are comprised in my BigML plan.

There are 8 tasks that, from the bottom to the top, correspond to:

- The upload/creation of source data from the CSV file.
- The creation of a dataset from this source (where I specified how to treat each feature — numeric, categorical, text).
- The creation of a model from the whole dataset (just out of interest).
- The removal of that model.
- The creation of the training set from a 50/50 split of the dataset.
- The creation of the test set.
- The creation of a model from the training set.
- The evaluation of that model on the test set.

You'll notice that what took the longest was splitting the data into training and test sets! At 1.445s for 2,354 predictions, this is around 600 nanoseconds for a single prediction on average, which is blazingly fast.

The first thing to look at after the training is the Decision Tree model itself, and features' importance as analyzed by BigML's training algorithms. This is also called "field importance" and is given in the "model summary":

```
Data distribution:
• False: 91.21% (2147 instances)
• True: 8.79% (207 instances)

Predicted distribution:
• False: 91.16% (2146 instances)
• True: 8.84% (208 instances)

Field importance:
1.   in address book: 33.11%
2.   content: 20.04%
3.   labels: 9.29%
4.   sender: 6.81%
5.   length: 6.53%
6.   %starred: 4.44%
7.   subject: 4.41%
8.   me: 3.98%
9.   time me reply: 3.34%
10.  attachments: 2.14%
11.  date.day-of-month: 1.86%
12.  conversations: 1.65%
13.  date.day-of-week: 1.52%
14.  date.second: 0.59%
15.  date.minute: 0.30%
16.  %read: 0.00%
17.  date.hour: 0.00%
18.  date.month: 0.00%
```

We know that the second the email was sent has no impact on the email's importance. BigML picked up a small importance value for this field/feature (0.59%). When a model finds relationships between features and outputs that do not exist, it is said to "overfit". All models overfit to a certain degree. If we see field/feature importance values that we would deem as too high, it would be an indicator of important overfitting and thus of poor model quality.

Bear in mind that this "field importance" is relative to the set of features that are being considered, so the values would be completely different if we would remove or add features. As a proof, here is what you get when removing all content-related features:

```
1.   % me replied: 47.48%
2.   %read: 10.62%
3.   length: 9.80%
4.   date.day-of-month: 5.84%
5.   %starred: 5.61%
6.   me: 5.35%
7.   length contact reply: 4.41%
8.   date.hour: 3.15%
9.   %from: 2.53%
10.  date.day-of-week: 2.33%
11.  in address book: 1.17%
12.  % contact replied: 0.74%
13.  conversations: 0.74%
14.  attachments: 0.25%
```

Some features (such as '% me replied' and '% read') now have much higher importance values. Note that these field/feature importance values correspond to one particular model, which in the end is just one way to interpret the data — it's not the truth. As George E. P. Box said, "All models are wrong, but some are useful." If your model turns out to be useful (based on performance measures), then you can indeed consider features that have high important values to have predictive power, but that doesn't necessarily mean the other features have none.

Let's now look at our model's performance on the test set (see next page's screenshot). With a Phi value of 0.70 for our model and 0 for algorithms with no intelligence whatsoever (saying that all emails are unimportant or making random predictions), we clearly see that our model is doing something smart. I also created a public version of a model without content features. I invite you to browse the Decision Tree[58], the corresponding dataset[59], and to have a look at the evaluation[60] of that model. Its F-measure is 0.81 and its Phi value is 0.63.

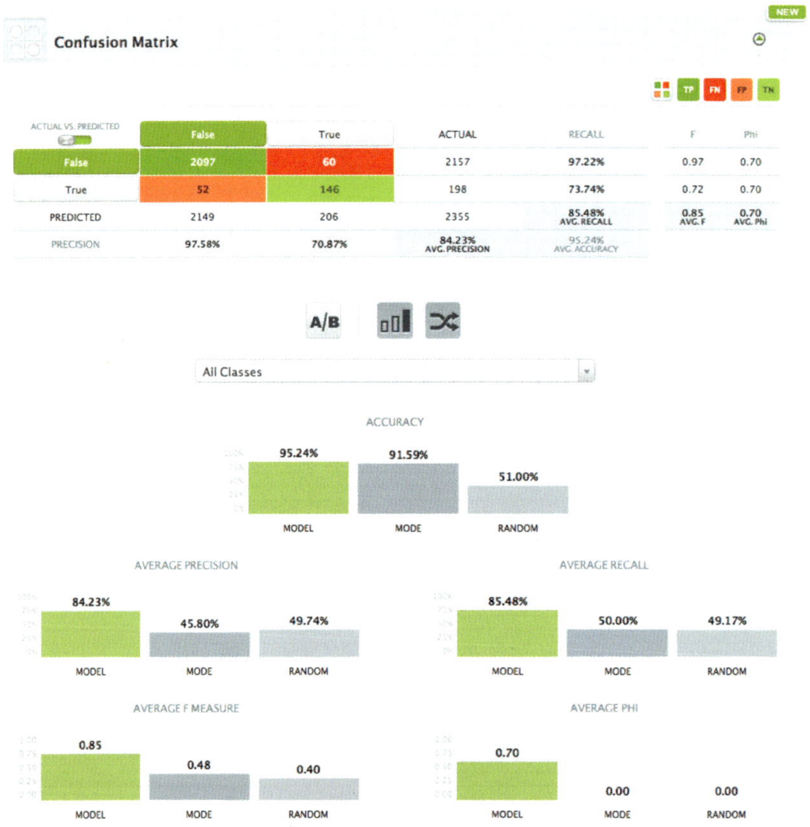

Our initial objective was 1 False Positive (FP) per day and 1 False Negative (FN) per 3 days on average. 2355 emails with 30 emails per day makes 78 days. Our objective was thus 47 FPs and 16 FNs, whereas we did 60 FPs and 52 FNs. The numbers of FPs and FNs are quite similar. It is likely that, in the near future, Prediction APIs will offer the option to penalize some errors more than others. But even if we had communicated to BigML that FNs were less desirable than FPs (so that we would have had less FNs), the number of FPs would probably have increased as a result (no free lunch!) and we would still have failed to reach our initial objectives.

The first idea that comes to mind to get better results is to fix the erroneous output values that we noticed while examining the data, and remove the noise as much as possible. We could try to add more training data by collecting more emails from the past, but we should keep in mind that the past, if it is too remote, may not be representative of the present. My guess would be that 6 months is okay, whereas the training data only spanned 2.5 months. Adding features would certainly help as well. For instance, we'll take the Google Plus Circles in which the sender is into account, or any information about the sender that can be collected from the Internet. (Is the sender an actual person? What's his Klout score?) If we had access to the mail client code, we could also measure the average time taken to read emails sent by the sender.

Our results could be affected by erroneous values in our data, or a lack of useful features. In general, the best way to find out more about this is to inspect the incorrect predictions and the predictions with low confidence values. Here in particular, we should inspect FNs to try to understand where the model has difficulties. Hopefully this would give us further insights into characterizing inputs in a way that would discriminate between positives and negatives.

CHAPTER 8: WRAP-UP

8

CONCLUSIONS

Back in May 2011, the highly influential management consulting firm McKinsey & Company published a white paper on Big Data. It predicted a shortage of people with expertise in Machine Learning. At the same time, Google Prediction API was released to the public, followed a few months later by BigML. I believe that ML-as-a-Service is the key to resolving this talent shortage issue. Sure, there are some ML problems that are not yet tackled by Prediction APIs, but it's only a matter of time before their capabilities expand.

ML is being democratized, and now that everyone can have easy access to it, we need to use it to create value at large. Hopefully this book will inspire you with new applications of ML to a domain in which you're an expert. Use your domain knowledge to get the best data, extract the best features, and find the most pertinent ways to use predictions. Also keep in mind that, as with any powerful new technology, certain uses can turn out to be harmful. In the case of ML, I think about certain forms of marketing and targeting. If you care about the implications of what you're doing, you should always question its nature, think about the long term, and think about people first (users and customers).

I am extremely excited by the wave of innovation that will spur from people like you using Prediction APIs, and I wish you a warm welcome to the community of Machine Learning bootstrappers!

RESOURCES TO DIG DEEPER

The topic of Machine Learning is vast. In this book, I chose to only focus on the most common types of ML problems, classification and regression. There are other types of problems and other types of ML — what you've read about here is referred to as "supervised learning". Below are a few pointers to resources of interest if you ever feel like learning more about ML:

- If you are interested in a business perspective, I recommend reading Getting Started with Business Analytics[61], or Data Science for Business[62] for a more comprehensive resource.

- To find out how ML algorithms work and gain expertise in the field, I recommend following Andrew Ng's Stanford course on Coursera[63].

- If you want to start building and running your own ML systems, I recommend learning Python[64] as a programming language, using Scikit-learn[65] as an ML library, and reading Programming Collective Intelligence[66] and Building Machine Learning Systems with Python[67].

ABOUT LOUIS DORARD

I studied Machine Learning at University College London, where I obtained my Ph.D. At the same time, I also bootstrapped an online business.

In between that and writing this book, I organized a challenge and workshop on website optimization at the International Conference on Machine Learning, and I then served as Chief Science Officer in a startup where I was in charge of starting a Research & Development program and turning it into a product.

I love starting things from scratch and I am passionate about technological innovation in web and mobile apps. My experience has taught me the importance of simplicity and of starting small.

My goal now is to help people create smarter apps and businesses, and train them to use Prediction APIs. For those who need more personalized help, I have created a consultancy, Codole[68].

You can find out more about me on LinkedIn (www.linkedin.com/in/louisdorard) and follow me on Twitter for updates on what I am working on (@louisdorard). If you have any thoughts to share, please email me at louis@dorard.me (in all likelihood your email will be detected as important by the Machine that receives it ;-)) and I will reply to you personally. I will also take your feedback into account in future revisions of the book.

ACKNOWLEDGMENTS

When tracking back how I came to write this book, the first person who comes to mind is Erick Alphonse, who I want to thank especially for introducing me to Prediction APIs and his support. The idea of writing a book itself was inspired by work from Nathan Barry and Jarrod Drysdale.

Even though writing this book was mostly a one-person adventure, I have received a lot of help along the way. Some of the content was guided by interesting discussions with fellow Machine Learners (hi Francisco Martin, Rand Hindi, Gabriel Synnaeve, Tom Diethe and Erick!), but also by people who were just interested in what I had to say and happy to share their thoughts. I am grateful to my early reviewers (Jean-Baptiste Goulain, Loïc Chauvet, David Bruant) and to my first readers for their valuable and extensive feedback, and to everyone who helped me spread the word about this book. That includes Camille Blaise, who's well on his way to become a social media guru, and who played a key role in the case study of Chapter 7 by introducing me to Google Apps Script! By the way, thanks to Alex Vallette too for showing me how to use Pandas!

Finally, thank you friends and family for all the support you gave me.

Oh, yes... and my girl, Adeline, qui m'a encouragé à me lancer dans cette aventure, tout en m'en distrayant, tout en étant la clé à un état d'esprit idéal pour aborder un travail de si longue haleine.

Merci chérie!

BORING NOTICE

Subsequent editions are not currently planned, but the author reserves the right to release them as separate paid products or as part of training courses, training kits, or other product formats.

NOTES

1. http://blog.bigml.com/2012/12/11/can-i-kick-it-predicting-kickstarter-success/

2. http://gmailblog.blogspot.fr/2010/12/priority-inbox-update-time-savings-new.html

3. http://snips.net/showcase.html

4. https://developers.google.com/transit/gtfs/

5. https://itunes.apple.com/fr/app/tranquilien/id660721122?mt=8

6. http://www.zillow.com/

7. http://www.redfin.com/

8. http://aws.amazon.com/mturk/

9. http://datamarket.azure.com/browse/data

10. http://www.wkiri.com/research/papers/wagstaff-MLmatters-12.pdf

11. http://pandas.pydata.org

12. http://pandas.pydata.org/pandas-docs/stable/10min.html

13. http://vimeo.com/59324550

14. https://wilwheaton.net/2013/12/goddamit-popular-science/

15. http://en.wikipedia.org/wiki/Recall_(information_retrieval)

16. http://scikit-learn.org/stable/modules/classes.html#classification-metrics

17. http://scikit-learn.org/stable/modules/cross_validation.html#random-permutations-cross-validation-a-k-a-shuffle-split

18. http://blog.mashape.com/post/48074869493/list-of-40-machine-learning-apis

19. http://www.alchemyapi.com

20. http://datumbox.com

21. http://semantria.com

22. http://www.maluuba.com/

23. https://www.mashape.com/nsure-io/porn-filter

24. http://blog.bigml.com/2013/07/01/you-dont-need-coursera-to-get-started-with-machine-learning/

25. https://bigml.com/user/francisco/gallery/model/52c1c3960c0b5e6fc900016d

26. http://www.bigml.com/

27. https://developers.google.com/prediction/

28. http://datamarket.azure.com/

29. https://docs.google.com/

30. http://aws.amazon.com/s3/

31. http://www.freebase.com/

32. https://bigml.com/user/BigML/gallery/model/4f89c39b1552686458000038

33. https://bigml.com/gallery/models

34. https://developers.google.com/prediction/

35. https://developers.google.com/apps-script/

36. https://docs.google.com/spreadsheet/

37. https://cloud.google.com/console/

38. https://developers.google.com/apis-explorer/

39. http://www.nltk.org/

40. https://developers.google.com/storage/pricing

41. https://developers.google.com/prediction/pricing

42. http://www.uclassify.com

43. http://www.predictobot.com/howitworks.html

44. https://developers.google.com/apps-script/

45. https://developers.google.com/apps-script/reference/gmail/gmail-thread#isImportant()

46. http://gmailmeter.com

47. http://gmailblog.blogspot.fr/2012/04/know-your-gmail-stats-using-gmail-meter.html

48. https://archive.ics.uci.edu/ml/datasets/Spambase

49. https://developers.google.com/prediction/docs/developer-guide

50. http://static.googleusercontent.com/media/research.google.com/fr//pubs/archive/36955.pdf

51. https://support.google.com/mail/answer/186531?hl=en

52. http://lifehacker.com/5992588/gmails-priority-inbox-is-awesome-just-give-it-a-chance

53. https://developers.google.com/apps-script/reference/gmail/gmail-app#search(String,Integer,Integer)

54. https://developers.google.com/apps-script/reference/gmail/

55. https://developers.google.com/apps-script/reference/contacts/

56. https://script.google.com/dashboard

57. https://developers.google.com/apps-script/reference/script-db/?hl=en

58. http://bml.io/1fXj3B0

59. http://bml.io/1bmK0B9

60. https://bigml.com/shared/evaluation/1za57i7DYhg0jNJ3RDslNClR0Bf

61. http://www.amazon.com/dp/B00ENTKU5E

62. http://shop.oreilly.com/product/0636920028918.do

63. https://www.coursera.org/course/ml

64. https://www.wakari.io/nb/url///wakari.io/static/notebooks/Lecture_1_Introduction_to_Python_Programming.ipynb

65. http://scikit-learn.org/

66. http://shop.oreilly.com/product/9780596529321.do

67. http://www.packtpub.com/building-machine-learning-systems-with-python/book

68. http://www.codole.com/

20172609R00115

Printed in Great Britain
by Amazon